# Darwinian Dynamics

# Darwinian Dynamics

## EVOLUTIONARY TRANSITIONS
## IN FITNESS AND INDIVIDUALITY

*Richard E. Michod*

PRINCETON UNIVERSITY PRESS

PRINCETON, NEW JERSEY

Second printing, and first paperback printing, 2000

Paperback ISBN 0-691-05011-2

The Library of Congress has cataloged the cloth edition of this book as follows

Michod, Richard E.
Darwinian dynamics : evolutionary transitions in fitness and
individuality / Richard E. Michod.
p.   cm.
Includes bibliographical references and index.
1. Natural selection.   2. Adaptation (Biology)   I. Title.
QH375.M535   1999
ISBN 0-691-02699-8 (cl : alk. paper)
576.8'2—dc21   98-4166   CIP

This book has been composed in Times Roman

The paper used in this publication meets the minimum requirements of
ANSI/NISO Z39.48-1992 (R1997) (*Permanence of Paper*)

http://pup.princeton.edu

Printed in the United States of America
2   3   4   5   6   7   8   9   10

*TO MY DAUGHTERS KRISTIN AND KAYLEY*

---

# Contents

CONTENTS

# Preface

LIFE EXISTS as hierarchically nested levels of organization, in which higher-level units are composed of lower-level units (gene, chromosome, genome, cell, multicellular organism, society). How did this come about, and what are the implications of hierarchical organization for individuality and the meaning of fitness in evolutionary explanation? Cooperation among lower-level units is central to the emergence of new higher levels, because only cooperation can trade fitness from lower to higher levels. My book is concerned with the study of cooperation, and the principles that guide the emergence of higher levels of organization. I have tried to show that there is a common set of principles and problems that bind the study of levels of organization as disparate as the gene, the cell, the multicellular organism, and whole societies. I focus here on the early transitions in the history of life (from genes to networks of cooperating genes to that first individual, the cell) and on the transition from single-celled organisms to multicellular ones.

These are exciting times for the study of cooperation. In the past, the study of cooperation has usually received less attention than the other two forms of ecological interaction, competition and predation. In the past scholars have viewed cooperation to be of limited interest, of special relevance to certain groups of organisms to be sure—the social insects, birds, our own species, and our primate relatives—but not of general significance to life on earth. All that has changed with the study of evolutionary transitions and the emergence of new units of selection. What began as the study of animal social behavior some thirty-five years ago has now embraced the study of interactions at all biological levels. Instead of being viewed as a special characteristic clustered in certain groups of social animals, cooperation is now seen as the primary creative force behind ever greater levels of complexity and organization in all of biology.

The benefit of group living results from cooperative interactions among group members. To create new levels of selection and organization in evolution, cooperation must be promoted among lower-level units, while, at the same time, ways must be found of mitigating the inherent tendency of the lower-level units to compete with one another through frequency-dependent fitness effects. Because cooperation is usually costly to the fitness of individuals within the group, defecting mutants can arise and take over the group, in the process destroying cooperation and the very conditions that made their increase possible in the first place. Cooperation is a critical factor in the emergence of new units of selection precisely because it trades fitness from the lower level (its costs) for increased fitness at the group level (its benefits). In this way, cooperation can create new levels of fitness (see tables 3-1 and 5-1).

The study of evolutionary transitions is the study of the emergence of new levels of fitness. Fitness is the most fundamental and unique concept in all of biology—pretty much everything else in biology is chemistry and physics, or a remnant of history. Although fundamental, it is difficult to define fitness and to explicate its role in evolutionary explanation, especially its role in evolutionary transitions. Fitness is both a cause and effect of evolutionary transitions. When it is traded between evolutionary units during cooperative interactions new levels of fitness and individuality may emerge. I hope to explain in the following pages how fitness is constructed out of evolutionary and population processes.

Understanding the emergence of higher levels of organization in evolution requires a population genetics theory of interaction in a multilevel context. I have tried to provide such a theoretical framework, and to apply it to evolutionary transitions from the molecular level to the whole organism. Further, because the study of evolutionary transitions is basically the study of the emergence of new levels of fitness, I have sought to clarify the role of fitness in this theory and in evolutionary explanation in general.

I take a dynamical point of view on evolution and on fitness concepts. Evolution has no enduring products—even organisms are of only fleeting existence, each born unique because of sex, each soon to die. Accordingly, I argue that to be construed correctly, fitness should apply to the process of genetic change (much as R. A. Fisher and G. Price envisioned) and not to products of evolution, such as organisms. I investigate fitness from the molecular level up to the level of the whole organism. With the understanding made possible by recent developments in ecology, multilevel selection theory, and the origin of life, it is now possible to present a theory of fitness. In so doing, I give special attention to fitness levels and their origins and transitions in the evolutionary process. Finally, I consider the philosophical implications of my theory of fitness for explanation in biology.

Before Darwin, design was understood as a product of either the human mind or a Creator. A watch implies a watchmaker, so argued William Payley in 1802. Darwin changed all that. Darwin argued that the design apparent in life arises out of processes intrinsic to life, not from extrinsic forces. Differential birth and death rates within populations of organisms—when systematically related to features in the environment—explain the well-designed features of organisms. The human eye, the grasp of the tiger's paw, the match of the pollinator with the flower, even the human mind—all must arise out of the blindly mechanistic process Darwin called natural selection.

Sounds simple, almost too simple, for there is a lot of explaining to do. How does the theory of evolution actually go about explaining design? One concept is central—fitness. Fitness is what makes biology different. But what is fitness and where did it come from?

The great philosopher Karl Popper dismissed biology as not being science, because he thought its central doctrine, "survival of the fittest," was a tautology. Who are the fittest? he asked. Those who survive of course. Darwin's great principle becomes the tautology "survival of those who survive," much to the exasperation of evolutionary biologists who have little fear that their science can be reduced to empty truisms. Popper's challenge has been heard by a generation of philosophers of biology who have come to the rescue of evolutionary science. By and large, these defenses have followed Darwin's lead and viewed fitness as a property of organisms.

It is easy to be confused by organisms, fitted as they are with such wondrous designs. Organisms are born and they die. In between this birthing and dying they may have offspring, some more than others according to the traits they possess and their environment. But this birthing and dying of organisms is only a part of the selection process, for the organism does not exist in isolation. In many situations, especially during evolutionary transitions, other factors dominate and interfere with the individual as the maximizing agent. Of particular concern to the major transitions in evolution are the frequency-dependent fitness effects within populations that frustrate emergence of higher levels of organization. The organism is not a maximizing agent, but this does not mean that the organism is not a unit of selection, as Dawkins argues;[1] rather it means that there is more to natural selection than maximizing individual fitness. Organizational factors like multiple levels of selection, genetic factors like epistasis, linkage, and recombination, and ecological factors like population density, frequency, and age structure can intervene and decouple organism fitness from evolution. I argue that a dynamically sufficient concept of fitness cannot relate to an overall property of organisms but instead must be associated with the dynamics of evolutionary change. The modern theory of evolution demands this dynamical view of fitness.

I develop a formal theory for the evolution of interactions using the dynamics of natural selection. Of course, the founding fathers of evolutionary

biology (Fisher, Wright, and Haldane) developed a basic selection theory—
so what has changed? There have been three key developments in the last
thirty or so years that make possible the synthesis I wish to make. First, the
abstract selection coefficient of the old theory has been unpacked and con-
nected to ecology and behavior, especially with regard to the role of interactions in affecting fitness. In Fisher's view, as embodied in his fundamental theorem of natural selection, the environment only deteriorates in time,
undermining the state of adaptation of the population. What was once a
black box, the environment, has now been given explicit ecological content.
Population biologists and ecologists understand the environment far better
and represent it explicitly in their models. Of central importance to evolutionary transitions is the role of interactions with other members of the population. Second, we understand in far greater detail the multilevel nature of
selection and how natural selection occurs simultaneously at different hierarchical levels: gene, chromosome, cell, organism, group, and species. As
a result of this multilevel approach, cooperation among different levels becomes not only possible but likely. I use the covariance approach to selection to embrace this multilevel setting. Finally, we have a sufficient (in principle) theory of the origin of life and genetic information. We understand
where fitness comes from and how it emerged from chemistry and physics.

Armed with these recent developments and a dynamical theory of selection, I reconsider Popper's challenge and its fallout in the philosophical literature. How does evolutionary theory explain design and the fitness and
individuality of evolutionary units?

# Acknowledgments

I WISH to understand in a common theoretical framework evolutionary transitions between levels of organization in biology. I have been influenced by the research of many workers, most significantly R. A. Fisher, W. D. Hamilton, J. Maynard Smith, E. O. Wilson, and S. Wright. I have also been influenced by three previous works on evolutionary transitions (E. O. Wilson 1975; Buss 1987; Maynard Smith and Szathmáry 1995). I learned from Buss (1987) that the principles of conflict and cooperation can be applied to the sorting out of cell lineages during development. I read Wilson's book (1975) as a graduate student, and, although it focused primarily on the transition from multicellular organisms to societies, I appreciated its far-reaching and synthetic approach. My earlier interest in the theory of kin selection and the evolution of sociality in both organisms and molecular replicators (Michod 1984, 1980, 1979, 1983b, 1982; Michod and Anderson 1979; Michod and Hamilton 1980) stemmed from the excitement I felt from reading Wilson's book (1975) and from visiting with John Maynard Smith and the rest of the Population Biology Group at the University of Sussex in 1976. I am especially indebted to John Maynard Smith, and will always cherish the many discussions we have had together.

Over the years I have had many discussions with many people about the topics considered in this book, including W. Anderson, H. Bernstein, R. Brandon, H. Byerly, R. Ferriere, P. H. Gouyon, F. Hopf, S. Kleiner, C. Lavigne, J. Maynard Smith, D. Roze, D. S. Wilson, E. Szathmáry, and B. Wimsatt. The following colleagues commented on the manuscript and their criticisms and suggestions greatly improved it: A. Ariew, H. Bernstein, R. Brandon, H. Byerly, S. Gavrilets, C. Lavigne, D. Roze, and M. Wade. I appreciate the discussions and comments of the participants in my seminar on fitness in spring 1997 at the University of Arizona (H. Byerly, A. Ariew, M. Clauss, A. Corl, E. Hebets, M. McIntosh, J. Netting, J. Lie, J. Reiss). I thank D. Roze and J. Lie for allowing me to refer to their unpublished results. I appreciate the hospitality of E. Szathmáry and the staff at the Collegium Budapest/Institute for Advanced Study, and R. Ferrier and the Ecole Normale Superieure de Paris, where I completed the book during my sabbatical in 1997–98. Grants from the National Science Foundation and the National Institute of Health funded the research upon which this book is based.

# Darwinian Dynamics

# The Language of Selection

## PLAN OF THE BOOK

I have written this book for two kinds of readers, those interested in the theory of evolutionary transitions (to new levels of fitness and individuality) and those interested in the role of fitness in evolutionary explanation. I realize that the evolutionary biologist may find the excursions into philosophy tangential, while the philosopher interested in the nature of biological explanations may find the emphasis on formal mathematical models overly technical. I ask the indulgence of both types of readers to tolerate the needs of the other.

In chapter 1, I provide a general introduction to the basic terms and concepts of selection theory used in the book. I am more interested in the meaning of a few central terms and concepts and their use in evolutionary explanation than in providing an exhaustive compendium. A more formal introduction to the mathematics of selection theory as it relates to the evolution of interactions is given later in chapter 4.

Cooperative interactions are the basis of more inclusive evolutionary units. Because cooperation reduces the fitness of lower-level units, while increasing the fitness of the group, cooperation drives transitions to higher-level units. Defection, the antithesis of cooperation, is the bane of cooperative groups everywhere, because it is often favored within the group by its frequency-dependent advantages. In light of the central role of cooperation in evolutionary transitions, I pay special attention to the means by which cooperative interactions evolve and are later enforced during the emergence of individuality.

In chapters 2 and 3, I study the early evolutionary transitions, from chemical compounds to replicating molecules to gene networks, culminating in that first true individual, the cell. I emphasize the role and meaning of fitness during these transitions and how the dynamics of natural selection of these early evolutionary units may be studied using simple models from population biology. During an evolutionary transition, frequency-dependent selection within the group prevents the maximization of the fitness of the group. For this reason, simple fitness arguments cannot be used in the study of emergent complexity during evolutionary transitions. I find that

3

there is much more to the dynamics of natural selection than increase of better-designed individuals. The central problem during evolutionary transitions is how to move from groups of lower-level units to a new level of selection. In short, the group must become an individual. But how does this occur? Conflict mediation is the process by which lower-level change is modulated in favor of the new emerging unit.

In chapters 5 and 6, I study the transition from cells to multicellular organisms. Mutation and selection among cells during development provide variation both within and between the cell-groups, or proto-organisms. However, the level of cooperation and synergism attained is rather limited, even in genetically related cell-groups clonally derived from a single cell, the zygote. Within-organism mutation and selection frustrate the creation of the new higher-level unit. The evolution of greater levels of cooperation among cells, the proximate agent of design at the organism level, is frustrated by within-organism change. Individual organisms are more than groups of genetically related cooperating cells; they require higher-level functions that mediate conflict within and maintain harmony for the group. Conflict mediators evolve to modify the rules of interaction and selection at the cell and group levels. Conflict mediators facilitate the transition to the new level by increasing the covariance between fitness at the cell-group level and genes carried in the zygote, which increases the heritability of fitness at the new higher level.

In chapter 7, I begin by considering the meaning of fitness that emerges from my study of evolutionary transitions. The concept of design in biology must rest on the Darwinian principles underlying natural selection, because there can be no preexisting plan in evolution. When the consequences of Darwin's principles are fully understood as embodied in the dynamics of natural selection, there is no meaningful concept of the individual as a maximizing agent, especially when the units of selection are changing. Frequency-dependent fitness effects frustrate the creation of new evolutionary units and prevent the application of simple maximization arguments. The tension between lower and higher level units is never completely resolved in any evolutionary transition. Units of selection, such as genes, cells, or organisms, do not exist in isolation, nor are they completely interdependent. Instead, they exist in a hierarchy of nested but partially decoupled levels, and any focal level provides the context for lower-level units as well as the components for higher-level units. Because evolutionary units (genes included) play the roles of both context and component at the same time, the dynamics of natural selection at any level involves an interplay between the dynamics at all levels.

After considering the meaning of fitness that emerges from my studies, I consider touchstone cases of natural selection using my dynamical view of fitness: the so-called tautology problem, the evolution of sex and reproductive systems, the immortality of the germ line in contrast to the mortality of the soma, such old standbys as sickle-cell anemia and heterozygote superiority, and the existence of species as distinct entities. These studies of natural selection recommend a dynamical view of fitness. This conclusion is pursued in more philosophical vein in chapter 8, where I consider other interpretations of fitness and offer a general framework for the role of fitness in evolutionary explanation.

## DARWINIAN DYNAMICS

Darwin proposed that evolution by natural selection occurs when a population of replicating entities possesses three characteristics: variation, heritability, and the "struggle to survive" (Darwin 1859). Darwin observed that all organisms have the potential to increase in numbers at an exponential rate and produce many more offspring than could possibly survive. Consequently, there must often be a competition in which some kinds of organism are better able to survive to reproduce. Such variations in survival and reproductive success (variations in individual fitness) provide the raw material for evolution by natural selection—so long as the traits are heritable (capable of being passed from parent to offspring). Darwin knew very little about how genes work, but he did appreciate that the tendency of offspring to resemble their parents is necessary for evolution. Today, we know that heritability results from the faithful, highly accurate replication of the nucleotide sequence information contained in DNA. Offspring resemble their parents because they possess copies of their parents' genes.

Putting together Darwin's properties of variation, the struggle to survive and reproduce, and heritability, we expect (Darwin certainly did) that heritable traits that aid in the struggle to survive and reproduce should increase in frequency in the population, and deleterious traits should decrease.[1] This is how natural selection is supposed to work. However, this expectation assumes that the unit of selection, usually taken to be the organism, is well established. During evolutionary transitions this is certainly not the case, and even after a new level has been established, lower-level change can produce results different than what we might expect. The view that more adapted traits always increase ignores many of the complicating factors that intervene during the life cycle of the organism—within-organism change, sex, genetic factors such as linkage and pleiotropy, and ecological factors such as population frequency and density. This book should make it clear that there are many more outcomes contained in Darwin's conditions and the process of natural selection than simply the increase of more fit organisms.

My colleagues and I term dynamical equations based on Darwin's conditions "Darwinian dynamics" (Bernstein et al. 1983; Byerly and Michod 1991a,b; Michod 1986, 1984). Darwinian dynamics deserve special study as a general class of equations with special dynamical properties that may be used to understand the process of natural selection. In addition to their fundamental role in biology, Darwinian dynamics occur in the study of ordered phenomena in certain physical systems such as lasers. Darwinian dynamics provide an explicit well-defined context for understanding the role of fitness in evolutionary explanations, as will be shown in the last two chapters.

Many definitions of natural selection have been offered. I view natural selection as the process of change in gene and genotype frequencies resulting from Darwin's conditions. In offering this definition, my emphasis is on the conditions involved in the process rather than the outcome. Indeed, I think there are several, perhaps many, possible outcomes of natural selection consistent with Darwin's defining conditions.

Darwin's conditions of variation and heritability in fitness can, in principle, apply at any level in the hierarchy of life: genes, gene networks, chromosomes, bacteria-like cells, eukaryote-like cells, multicellular organisms, families, groups, and societies (Lewontin 1970). In this book, I am particularly interested in the progression of fitness relations as new levels of selection are created during evolution. The essence of a transition in evolutionary units is that lower-level units relinquish their claim to fitness, as it were, so that fitness may emerge at the new higher level. Cooperation drives the passage from one level of fitness to another, because cooperation trades increased fitness at the higher level for decreased fitness at the lower level. In chapters 2, 3, 5, and 6, I consider the conditions under which evolutionary transitions may occur. A more formal presentation of Darwin's conditions is given in chapter 8, in the section "Natural Selection as a Biological Law."

The orderly transformation of living systems through time depends on information encoded in the sequence of nucleotide base pairs in DNA. Genetic information must accomplish two seemingly contradictory feats, which baffled early geneticists: while accurately transferring itself from parent to offspring, the genetic material must be open to modification in the event that the environment changes. Informational noise—random changes in the information sequence—is both a curse and a source of hope in this endeavor. Ultimately, all the variation underlying design in the living world comes from random changes to DNA, through either mutation or recombination.

Processes that produce informational noise should generally reduce fitness unless they also produce a benefit that offsets this deleterious effect. In living systems, informational noise is produced during the repair and replication of DNA (Bernstein et al. 1987)—the benefits of DNA repair and replication are self-evident. I have discussed elsewhere how recombination copes with genetic error while maintaining an open and flexible genetic system (Michod 1995). For natural selection to occur, the information in the genes must be perpetuated through time, or else the properties and relations of living things would not be heritable, that is, passed on from parent to offspring. Without the heritability of traits contributing to fitness, the life cycle would crumble, since offspring would no longer resemble their parents, and the flow of information underlying the fitness relations so necessary for the evolution of life would cease. In chapter 6 I pay special attention to how heritability of fitness may emerge at a new level.

## MAJOR EVOLUTIONARY TRANSITIONS

Natural selection requires heritable variations in fitness. Levels in the biological hierarchy—genes, chromosomes, cells, organisms, kin groups, groups, societies—possess these properties to varying degrees, according to which they may function as units of selection in the evolutionary process (Lewontin 1970). From E. O. Wilson (1975) and the transition from solitary animals to societies, to Buss (1987) with the transition from unicellular to multicellular organisms, and more recently Maynard Smith and Szathmáry (1995; Szathmáry and Maynard Smith 1995), attention has focused on understanding transitions between different levels of selection.

The major transitions in evolutionary units are from individual genes to networks of genes, from gene networks to bacteria-like cells, from bacteria-like cells to eukaryotic cells with organelles, from cells to multicellular organisms, and from solitary organisms to societies (Buss 1987; Maynard Smith and Szathmáry 1995; Maynard Smith 1991, 1990, 1988). These transitions in the units of selection share two common themes: the emergence of cooperation among the lower-level units in the functioning of the new higher-level unit, and regulation of conflict among the lower-level units.

Eigen and Schuster proposed the hypercycle as a way to keep individual

7

genes from competing with one another so that cooperating gene networks could emerge (Eigen and Schuster 1979; Eigen 1992). Localizing genes in the cell keeps selfish parasitic genes from destroying the cooperative nature of the genome (Michod 1983b; Eigen 1992; Maynard Smith and Szathmáry 1995). These early transitions from genes to gene networks to cells are discussed in chapters 2 and 3. Chromosomes reduce the conflict among individual genes (Maynard Smith and Szathmáry 1993, 1995). Meiosis serves to police the selfish tendencies of genes and usually insures that each of the alleles at every diploid locus has an equal chance of ending up in a gamete. As a result of the fairness of meiosis, genes can increase their representation in the next generation only by cooperating with other genes to help make a better organism. Uniparental inheritance of cytoplasm may serve as a means of reducing conflict among organelles through the expression of either nuclear genes (Hoekstra 1990; Hurst 1990; Hastings 1992), or organelle genes (Godelle and Reboud 1995), or both. Finally, concerning the final transition—that from organisms to societies of cooperating organisms—the theories of kin selection, reciprocation, and group selection (introduced in chapter 4) provide three related mechanisms for the regulation of conflict among organisms: genetic relatedness, repeated encounters, and group structure. These are just a few of the ways in which the selfish tendencies of lower-level units are regulated during the emergence of a new higher-level unit.

As initially conceived, the field of sociobiology focused on the transition from solitary organisms to groups of organisms, or societies, and the emergence of cooperative functions at the social level, the level of the colony, say, in the case of eusocial behavior in insects (E. O. Wilson, 1975). However, the set of tools and concepts used in studying conflict and cooperation during the transition from organisms to societies has proved useful for studying the other major transitions.

## COOPERATION AND CONFLICT

New evolutionary units begin as groups of existing units. Two issues are central to the creation of a new unit of selection: promoting cooperation among the lower-level units in the functioning of the group, while at the same time mitigating the inherent tendency of the lower-level units to compete with one another through frequency-dependent fitness effects. Cooperation represents the benefit of group living; groups of individuals can then behave in new and useful ways. Cooperation is a critical factor in the emergence of new units of selection precisely because it trades fitness at the lower level (its costs) for increased fitness at the group level (its benefits). In this way, cooperation can create new levels of fitness (see, for example, tables 3-1 and 5-1).

Frequency-dependent interactions among evolutionary units are both a source of novelty for the group and a threat to its collective well-being. Because cooperation is usually costly to the fitness of the individuals involved, defection may reap the benefits of cooperation and spread in the population, thereby destroying the very conditions upon which its spread depended in the first place. As a result of the spread of defection, cooperation is lost and along with it any hope for the creation of a new higher level.

Certain conditions are required to overcome the inherent limits posed by frequency-dependent selection to the emergence of new levels of selection: kinship, population structure, and conflict mediation. Conflict mediation is the process by which lower-level change is modulated in favor of the new emerging unit.

The theory of cooperation and conflict presented here is concerned with populations of interacting and replicating entities (genes, cells, organisms) that share a common ecological and/or genetic context. For the most part, this means either a common resource base in the case of replicating molecules studied in chapters 2 and 3 or a common gene pool and group setting in the cases of cells within organisms studied in chapters 5 and 6. I study symbiosis (cooperation between genetic units that were once capable of independent existence) in the context of the origin of cooperating networks of genes in chapter 3, in the game theory of cooperation in chapter 4, and in the transition from cells to organisms in chapters 5 and 6. However, I do not study the symbiotic origin of mitochondria, chloroplasts, and microbodies that make up the eukaryotic cell (Margulis 1970, 1981). Symbiosis is of fundamental importance to the emergence of complexity during evolution, but there are also other ways in which evolutionary transitions may occur (Maynard Smith and Szathmáry 1995).

## FISHERIAN FITNESS

There are many legitimate notions of "fitness." In discrete-generation population genetics models, fitness is often defined as the expected reproductive success of a type (reproductive output weighted by survival). I refer to this notion of fitness as individual fitness.[2] There are other meanings to the term "fitness," however. In his explorations into natural selection, R. A.

9

Fisher defined fitness as the rate of increase of a type, often expressed on a per capita basis (1930). I refer to Fisher's notion of fitness as *F-fitness*. It is difficult to commit to any one definition of fitness, because there are many legitimate definitions, each suited to a different purpose. On this topic, I have much more to say. However, at the outset let me say that I find Fisher's notion of fitness by far the most encompassing and dynamically sufficient concept of fitness, especially when the levels of selection are changing. For this reason, I return to Fisher's concept when discussing philosophical matters in the last two chapters.

Why did Fisher focus on the rate of increase? Because it "measures fitness by the objective fact of representation in future generations" (Fisher 1958, p. 37). The per capita rate of increase is the bottom line in population studies in both ecology and genetics. For a discussion of the central role of the per capita rate of increase in both population ecology and population genetics, see Ginzburg 1983, especially chapter 1. For an especially clear presentation of *F*-fitness in the context of genetic models, see Denniston 1978 (where *F*-fitness is termed $r_{ij}$). Fisherian fitness is given in equation (1-1), letting $X_i$ denote the density or frequency of type $i$ in a population and $t$ denote time:

$$F_i \equiv \frac{1}{X_i} \frac{dX_i}{dt}. \tag{1-1}$$

A genotype's rate of increase expresses its reproductive success over time and ultimately determines its evolutionary success (at least for the short term, which tends to be how natural selection acts). Confusion may arise, however, because the rate of increase of a genotype does not usually depend on properties of the genotype in isolation. In addition to depending on the environment and the composition of the population, a genotype's rate of increase in a sexual population depends upon the properties of the genotypes of all potential mates.

Fisher referred to the rate of increase of a genotype as the "Malthusian parameter," denoted $m$. Later workers often represented $m$ as if it were a property of the genotype alone, expressing it in terms of the genotype specific rates of birth and death as follows (Crow and Kimura 1970):

$$\frac{1}{X_i} \frac{dX_i}{dt} = m_i \stackrel{?}{=} b_i - d_i \tag{1-2}$$

The problem with the decomposition of fitness given in equation 1-2 is that the per capita rate of increase cannot usually be expressed as a function of individual genotypic capacities alone, capacities such as $b_i$ or $d_i$. Because this is misleading as a general analysis of fitness, I have put a "?" above the equal sign in equation 1-2. Later work has shown that the formulation given in equation 1-2 assumes fixed per capita birth and death rates (Charlesworth 1980), assumptions that are unlikely to hold if the genotypic composition of the population is changing, especially if genotypes interact in their effects on fitness, or if there is sex. Almost everything going on in the population affects the rate of increase of a type, and so Fisherian fitness cannot be regarded as the sole property of the individual or genotype.

Nevertheless, as we see in later chapters, Fisher's term "Malthusian" is appropriate, because the decomposition of fitness embodied by the right-hand side of equation 1-2 holds in a meaningful sense, though not in the sense of the equality given in equation 1-2. When population growth is a linear function of density, exponential or Malthusian growth results. The goal embodied on the right-hand side of equation 1-2 is to express evolutionary success (the left-hand side of the equation) as a function of the individual capacities alone given on the right-hand side. When population growth is Malthusian (linear in density), the condition for mutant invasion involves only $b_i$ and $d_i$ (see equation 2-5 or equation 3-3 below). So the decomposition of evolutionary success suggested by equation 1-2 applies to mutant invasion (even though the equality in equation 1-2 does not generally hold). This is why I find Fisher's term "Malthusian" especially appropriate.

The approach embodied in equation 1-2 formulates a goal of the theory of natural selection—to express evolutionary success of a type, say a genotype, as a function of the genotype's individual capacities and characteristics. In equation 1-2, evolutionary success is represented by genotype $i$'s fitness in Fisher's sense as a rate of increase, $(1/X_i)$ $(dX_i/dX_i)$, and the genotype's capacities are represented simply by $b_i$ and $d_i$. More generally, the genotype's capacities can be any attributes of interest. In this respect, the decomposition of fitness studied in chapters 7 and 8 follows Fisher's lead. The challenge is to express evolutionary change as a function of individual capacities in a way that is dynamically sufficient and accurately represents the underlying causal nexus of individual, genetic, ecological, and population factors, something that equation 1-2 fails to do.

## DECONSTRUCTING FITNESS

Evolutionary explanations based on natural selection require the construction of models that involve the causal components of fitness for the problem or situation under study. Fitness must first be decomposed into its components, and then reconstructed in the context of predicting and understanding evolutionary dynamics. To understand fitness and its role in the evolutionary process, I take care to separate out the components of fitness—as motivated by the particular question at hand. Fitness decomposition involves defining the heritable capacities underlying activities that make up the life cycle and ultimately lead to survival and reproductive success. Almost all models of natural selection involve some kind of fitness decomposition in one form or another. The particular traits analyzed are suggested by the problem and questions under study. In the reconstruction of fitness, the heritable capacities are related to the appropriate environmental variables (such as density, frequency, or resources) along with aspects of the genetic and reproductive system to yield causal statements of the dynamics of selection.

Fitness depends on a great variety of ecological factors and conditions. It may depend upon the density of genotypes or the total size of the population, as in density-dependent selection (Anderson 1971; Asmussen 1983a,b; Asmussen and Feldman 1977; Asmussen 1979). Fitness may depend upon the age or life history stage of the individuals involved, as during age- or stage-specific selection. In age-specific selection, fitness is decomposed into genotype-specific life tables of fecundity and viability rates (Charlesworth 1980; Anderson and King 1970). Fitness often depends upon interactions with other types in a population, for example, in frequency-dependent selection. This is especially true when there are transitions in levels of selection fueled by cooperative interactions. Interactions may occur between members of the same species, as is the case with the evolution of cooperative or selfish social behaviors, or between members of other interacting species such as competitors, predators, or mutualists. In ecological models of interacting species, the fitnesses of the genotypes within one species may be modeled as Lotka–Volterra–like functions of the densities of genotypes in the other species complete with genotypic-specific intrinsic rates of increase, carrying capacities, and competition coefficients (Roughgarden 1979; Leon 1974; Leon and Charlesworth 1978). The genotype-specific coefficients in ecological models of selection are examples of what I call "heritable capacities."

Since the life cycle requires both survival and reproduction, the most basic distinction among traits is whether they affect survival, fertility (mating success), and/or fecundity (numbers of gametes or offspring produced). Some traits affect all three components of fitness. The fundamental distinction between birth and death is reflected in Fisher's Malthusian parameter (equation 1-2). Fertility selection is more complex than viability selection. Because of sex, fertility and fecundity depend upon the mating pair, a complication that is commonly ignored in mathematical models because of the accompanying increase in the complexity of the model. With just two alleles at a single locus there are three genotypes and nine possible mating pairs (seven if we don't distinguish the gender of the parent). This so-called dimensionality problem (Lewontin 1974) arises in large part because of sex. Darwin went so far as to view selection on traits and behaviors related to sex and mating as sufficiently different from viability selection that he distinguished sexual selection from "natural selection." By "natural selection" Darwin meant viability selection. Contemporary workers understand sexual selection and viability selection as components of natural selection.

There is more to the decomposition and reconstruction of fitness and the analysis of the fitness components. Fitness may also vary in time, space, and level of organization. In this book, I am interested in spatial and organizational structures and their roles in the evolution of interactions and the emergence of new levels of individuality. Organizational structure involves the way in which evolutionary units are packaged into higher-level units and the rules involved. Conflict mediation of lower-level units through adaptations such as germ lines and self-policing are studied in chapter 6. Population structure encompasses local interactions that may differ among regions of space. Spatial variations in interactions arise because of restricted mobility, which may in turn depend upon distance or physical barriers to movement. Population structure may give rise to groups of interacting individuals where the local fitnesses of individuals depend upon their group membership (group selection). Interactions among relatives may also occur, as in kin selection. Genetic relatives share genes identical by descent. When interactions occur among genetic relatives inclusive fitness guides the evolutionary process (see chapter 4).

## SELECTION AS FITNESS COVARIANCE

In common usage, "selection" means choosing a subset of objects (such as food items in a grocery store) according to their properties (perhaps their price). Natural selection is quite different (see table 1-1), and commonsense expectations based on a mistaken analogy between natural selection and commonplace selection lead only to confusion when trying to understand evolution by natural selection. Examples of such problems to be discussed in this book include Fisher's fundamental theorem (see chapter 4) and the

| Subset Selection | Natural Selection |
|---|---|
| Choosing agent | No choosing agent; instead differential survival and reproduction causes change |
| No property change of entities involved in selection | Property change common as offspring differ from parents due to within-level change, sexual recombination, mutation, and other aspects of genetic and reproductive systems |

*Table 1-1*

Differences between natural selection and commonplace subset selection.

view that "selection" can be separated from the process of natural selection (Brandon 1991).

Under natural selection there is no entity acting as a choosing agent. Instead, the "selected set" is generated by a variety of different processes, including not only differences in individual fitness (expected reproductive success) but also effects of the genetic and reproductive systems. The "selected set" contains not only a subset of the previous generation (if generations overlap) but also new offspring. Because of sexual recombination and forces of within-organism change (such as mutation during development), these offspring may have different properties from their parents. This is as if we selected grocery items according to low price and got home to find the most expensive items in our grocery bag. This "property change" does not usually occur in everyday subset selection, but property change is common in natural selection, where it acts with differential survival and reproduction to produce overall change in the population (Price 1995).

Because offspring involve both property change and "selection," it is often not possible to cleanly separate out the selecting or sorting step from the process of natural selection. Some philosophers have argued just the opposite, and furthermore, that "fitness" should be restricted to the "selection" or sorting step alone (Brandon 1991; Mills and Beatty 1979; Rosenberg 1985, 1983; Sober 1993). These matters will be taken up in the final two chapters.

Price (1995) has shown that it is possible to unify the different meanings of "selection" by focusing on the covariance between "fitness" and the selected properties of the objects or units of selection. I now provide the calculations leading to Price's central result, the celebrated covariance equation 1-7. Let us classify the objects of the pre- and post-selection sets or

14

populations by the index $i$. For each object before selection, consider its value for a property of interest, $q_i$, and its frequency (or some other measure of its abundance) in the pre- and post-selection populations, $f_i$ and $f'_i$, respectively. Selection may be said to occur when the frequencies after selection, $f'_i$, are nonrandomly related to the properties $q_i$ before selection. This can be expressed more precisely as follows (Price 1995). The average property in the population before selection, $\bar{q}$, is given by

$$\bar{q} = \sum_i q_i f_i, \tag{1-3}$$

and the average property after selection (assuming the properties of the objects do not change) is

$$\bar{q}' = \sum_i q_i f'_i. \tag{1-4}$$

Price defines fitness, $W_i$, as the ratio of the frequencies before and after selection,

$$W_i = \frac{f'_i}{f_i}. \tag{1-5}$$

Fitness as defined in equation 1-5 is the relative change in frequency, or some other measure of abundance, from one generation to the next. Its role in selection can be further decomposed. In equation 1-6, I consider the covariance of fitness, $W_i$, with the properties $q_i$ and simplify it into a form that is useful in studying the effect of selection on the change in the average properties of the population:

$$\text{Cov}[W_i, q_i] \equiv \sum_i f_i q_i W_i - \sum_i f_i q_i \sum_i f_i W_i$$

$$= \sum_i q_i f'_i - \bar{q}. \tag{1-6}$$

Using the last form of equation 1-6 and the average properties of the unselected (equation 1-3) and selected (equation 1-4) populations, the change in the average property of the population is given by

$$\Delta q = \bar{q}' - \bar{q} = \text{Cov}[W_i, q_i]. \tag{1-7}$$

Equation 1-7 states in mathematical form our intuitive notion of selection, which is simply that the change in frequency of objects after selection (more precisely, the ratio of frequencies after selection to before) is related to some property of the objects. The relevant relation is that of statistical covariance of fitness (equation 1-5) with a heritable property of the units involved ($q_i$). In chapters 5 and 6, I use an extension of equation 1-7, in which $W_i$ is the fitness of the adult organism and $q_i$ is the frequency of a gene in the zygote stage before development.

According to Price's general formulation of selection, fitness describes a transformation in frequencies through time (equation 1-5). This definition is similar to the definition of fitness proposed by Fisher (1958) and discussed above. Fisherian fitness is also the relative increase in frequency of a type expressed on a per capita basis. Fisher's and Price's approach to fitness figures prominently in my studies in this book. In the final two chapters, I explore the general use of fitness in evolutionary explanations and argue for a dynamical view of fitness based on its role in mediating life cycle transformations through time. Price's approach will be developed further in chapters 4, 5, and 6.

## Mathematical Models

I use simple mathematical models extensively in my studies in this book. Models show what is possible, based on assumptions about how the world works. They cannot by themselves prove that a hypothesis is true. However, they can rule out poorly formulated or illogical hypotheses as well as suggest new hypotheses and fruitful lines of inquiry. By guiding experiment and observation, models are an integral part of scientific discovery. I primarily use simple population genetics models because they have great predictive and heuristic value in the understanding of complex evolutionary dynamics (Michod 1986, 1981; Provine 1986, 1977, 1971; Ruse 1973; Wimsatt 1980).

It is essential that a field of inquiry has clear logical and conceptual relations among the terms and ideas it employs. Evolutionary biology has experienced a surge in terms and ideas concerning the evolution of interactions and the meaning of fitness when there are frequency-dependent interactions. Consider only a few: evolutionarily stable strategy, kin selection, group selection, parental manipulation, reciprocation, social selection, family selection, sexual selection, the handicap principle, and many others. This proliferation is a healthy shift from the early days of population genetics when the selection coefficient was an abstract construct, assumed constant and devoid of ecological content. Mathematical theory may help us to understand the relations among these terms and the concepts they entail. History shows that this use of theoretical population genetics leads to the integration of ideas and serves to guide experiment and observation in evolutionary biology (Byerly and Michod 1991a,b; Michod 1986, 1984, 1981; Provine 1977, 1971; Ruse 1988; Wimsatt 1980). Mathematical models also help to clarify certain distinctions that are difficult to make in purely verbal formulations. For this reason, I use mathematical models to help us to understand evolutionary explanations based on natural selection.

| Term or Concept | Definition or Usage |
| --- | --- |
| Self-replication | The capacity to make copies, so that even mistakes can be copied |
| Natural selection | The process of genetic change resulting from Darwin's three conditions of variation, heritability, and the struggle to survive and reproduce |
| Individual Fitness | Expected reproductive success of a type |
| Fisherian Fitness | Per capita rate of increase of a type |
| Heritable Capacity | A heritable trait or capacity that is a component of fitness and represents the contribution of the phenotype's design attributes to natural selection |
| Cooperation | An interaction that decreases the fitness of the individual but increases the fitness of the group |
| Frequency-Dependent Selection | When individual fitness depends upon interactions within a group or population |
| Individual | A unit of selection satisfying Darwin's three conditions with mechanisms to modulate lower-level change |
| Conflict | Competition among lower-level units of selection leading to defection and a disruption of the functioning of the group |
| Conflict Mediation | The process by which the fitnesses of lower-level units are aligned with the fitness of the group |
| Fitness Covariance | The covariance between individual fitness and heritable genetic properties (used in chapter 6 to study the emergence of individuality |

*Table 1-2*

Definitions of terms and concepts.

## Adequacy Criterion for Understanding Fitness

Much has been written about the meaning and use of fitness in evolutionary biology. In analyzing a concept so embedded in evolutionary thought, but so open to different (yet often legitimate) interpretations, it is difficult to know where to begin and what to assume. For this reason, I appreciate Rosenberg's approach of stating up front an adequacy criterion for understanding fitness (1991). I take the approach that we understand fitness when we know three things. First, when and how does fitness originate in the biological world during the transition from the nonliving to the living? Second, how do new levels of fitness emerge during the evolutionary process? Third, how does fitness enter into mathematical models of natural selection? When we understand the answers to these three questions, we will be able to understand the meaning of fitness and its role in evolutionary thought. In the next chapter, I consider the first of these questions.

## Definitions of Basic Concepts

The definition or usage of certain fundamental terms and concepts used in this book is given in table 1-2.

# Origin of Fitness

WHEN AND HOW did fitness originate during the transition from the physical world to the world of the living? Fossil evidence indicates that eukaryotic cells were in existence about two billion years ago. Less complex cells with gross morphological properties similar to those of modern blue-green algae were likely in existence around three billion years ago. This leaves about a billion and a half years for evolution to produce a cell from the utter chaos and confusion that must have pervaded the primitive earth. Workers make informed guesses concerning the practical details of this critical period for the origin of life, based on extant chemistries and inferred conditions of the primitive earth. What principles guided the origin of life and creation of the first cell? Fortunately, we have a better understanding of the steps and the likely properties of the players involved, even if we can only guess at the underlying chemistries. The key to the origin of life and fitness is the origin of both self-replication and the replication of error, that is, mutation.

## COMPLEMENTARITY

Replication of molecules may occur when their shapes and chemistry possess a property called "complementarity." As a result of complementarity, a molecule may serve as a template for the production of its copy. The role of complementarity in the replication of nucleic acids is well known and taught in elementary biology texts. Complementarity may also drive the self-replication of synthetic molecules (Rebek 1994).

I base my discussion of the origin of life on the single-strand RNA scenario considered by many workers in the field to be a likely paradigm for the origin of life (Cech 1986; Bernstein et al. 1983; Michod 1983b; Eigen and Schuster 1979; Eigen 1992; Schuster 1980; Eigen et al, 1981; Orgel 1987, 1986, 1992). Single-stranded RNA (ssRNA) is an especially likely candidate for the original genetic material, because it possesses both the capacity for self-replication and the catalytic properties of an enzyme, such as the ability to catalyze alkylation and highly specific hydrolytic and transesterification reactions present in self-splicing and self-cleaving (Cech 1987). Because it has both replicative and enzymatic properties, ssRNA neatly solves the chicken and egg dilemma concerning whether enzymes (the chicken) or nucleic acids (the egg) came first. Nevertheless, there remain practical problems with the theory (Maynard Smith and Szathmáry 1995; Kauffman 1993). I believe that the points made here do not specifically depend on the practical details of ssRNA or even nucleic acids. If a substance other than RNA or DNA composed the first replicators, it would have to have the capacities of self-replication and mutation to evolve and could serve as the basis for the processes and transitions that I envision.

## SPONTANEOUS CREATION

Short-length nucleotide sequences are formed under conditions thought to be reasonable for the primitive earth (Lohrmann and Orgel 1980; Lohrmann et al. 1980; Schuster 1980; Lohrmann 1975; Orgel 1987, 1986; Miller 1987). Let $X_i$ be the frequency of a specific nucleotide sequence formed spontaneously under prebiotic conditions. The dynamic describing the change in frequency of $X_i$ is given by

$$\dot{X}_i = \beta_i - \delta_i X_i. \tag{2-1}$$

This equation is aptly described as spontaneous creation since it represents growth from zero density. The parameter $\beta_i$ describes the rate of conversion of precursor molecules into molecules of type $i$ and is presumed to take into account the local densities of these substrates. Spontaneous creation is not a birth process since the production of type $i$ is independent of the density of $i$, $X_i$. Birth requires replication, which means that as the type density decreases to zero so must the type's rate of production. For spontaneous creation (equation 2-1) the rate of production of $i$ is constant and independent of type $i$'s density. For this reason, equation 2-1 is not a Darwinian dynamic. Although it may be tempting to think of $\beta_i$ and $\delta_i$ as describing birth and death, this would be incorrect. The parameters $\beta_i$ and $\delta_i$ describe the forward and backward rates of the same reaction and ultimately both depend upon the activation energy between the substrates and the oligonucleotide product (figure 2-1). These parameters do not describe functionally separate birth and death capacities. While we could view the persistence of $X_i$ as a measure of its fitness, there is no life history—and so there can be no fitness in a biological sense.

## SELF-REPLICATION AND THE ORIGIN OF FITNESS

Fitness emerges when molecules become capable of self-replication. I am not concerned here with the practical details of how this happened, which

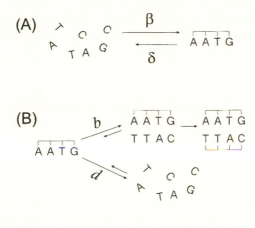

*Figure 2-1*

Spontaneous creation (A) and self-replication using a template (B). Adapted from Michod 1983b.

are considerable (Schuster 1980; Maynard Smith and Szathmáry 1995). Rather, I am interested in the consequences of self-replication for the dynamics of evolution and the emergence of fitness.

An important consequence of template-mediated self-replication is that the molecule may enter into two very different kinds of reactions (figure 2-1). What this means in terms of the population biology of the replicator is nothing short of profound, because the replicator now has a life history. The replicator may serve as a template for its own replication via the polymerization of a complementary strand or it may be degraded into its mononucleotide building blocks. Furthermore, different features of the molecule adapt it to undertaking these two reactions. A more compact molecule should be more stable to hydrolysis while a more open molecule would replicate faster, because an open structure would make it easier for nucleotide building blocks to find their complement in the template strand. Whether a molecule folds into a compact structure or not depends upon the nucleotide sequence of the molecule and the environment (pH, etc.).

## REPLICATOR DYNAMICS

Template-mediated replication (without the aid of enzymes) can be considered to involve two processes, although this is a simplification of many intermediate steps (Bernstein et al. 1983). First, the nucleotide template forms a template-resource complex with energy-rich mononucleotides. Second, this complex dissociates into two templates (parent and daughter strands). Using Michaelis–Menton kinetics the dynamic given in equation 2-2 was previously derived:

$$\frac{\dot{X}_i}{X_i} = \frac{b_i R}{R_i + R} - d_i \qquad (2.2)$$

(Bernstein et al. 1983). This is an example of a Darwinian dynamic. In equation 2-2, $R$ is the concentration of resources in the local environment and $X_i$ is the density of a particular nucleotide sequence $i$. The per capita rate of increase is given as a function of three heritable capacities discussed in table 2-1 ($b_i$, $d_i$, $R_i$) and the environment, $R$. Variation in the heritable capacities of the replicators is indexed by the subscript $i$. Heritability is represented by the fact that offspring molecules have similar capacities as parent molecules. Resources are taken to be mononucleotide building blocks available through prebiotic synthesis and from decomposition of dead replicators.

Depending on how template replication and resource utilization and limitation are modeled, other replicator dynamics are possible. For example, Eigen and coworkers assume that the total number of replicators is maintained constant by subtracting the total productivity of the replicator community (Eigen et al. 1981; Eigen and Schuster 1979, 1978a,b, 1977). I use Eigen's formulation below when I consider the transition from simple replicators to networks of cooperating replicators (hypercycles) and the evolution of proteins and protocells. Another possibility is given by Szathmáry (1991), in which birth (replication) is modeled as a square root function of density. The way in which density affects replication dramatically affects the outcome of natural selection (Szathmáry 1991; Michod 1991, 1983b; Eigen and Schuster 1979), leading to the different selection paradigms discussed later in the chapter.

## DESIGN ANALYSIS OF MOLECULAR REPLICATOR

The sequence of nucleotides present in the template molecule constitutes its genetic information or genotype, and the three-dimensional structure of the molecule its phenotype. At this early stage, without the presence of proteins encoded by the template, the phenotype and genotype were inextricably linked. Without enzymes there is no promotion of birth and avoidance of death through protein products encoded in the genotype. The phenotype is simply those properties of the replicator molecule that determine the heritable capacities given in table 2-1. Only with the evolution of a protein code could the phenotype and genotype diverge (Michod 1983b).

| Capacity | Definition | Meaning |
| --- | --- | --- |
| $b_i$ | Maximum per capita birth rate under conditions of excess resources, $R >> R_i$ | Heritable capacity at birth |
| $R_i$ | Concentration of resources at which the rate of birth is half maximum, analogous to Michaelis constant | Heritable capacity at utilizing resources |
| $d_i$ | Rate of dissociation of replicator into resources | Heritable capacity at avoiding death |

*Table 2-1*

Heritable capacities of the molecular replicator.

What structural features of the molecule determine the heritable capacities given in table 2-1? My colleagues and I have discussed previously some of the issues involved in understanding the design and fitness properties of a molecular replicator (Bernstein et al. 1983; Michod 1983b). My presentation here closely follows these previous discussions. The building blocks, presumably ribonucleoside 5'-phosphates, were likely available through prebiotic synthesis (Lohrmann 1975). The maximum rate of replication, $b_i$, depends upon the rate of elongation of the new RNA chain in terms of nucleotides added per unit time, the frequency of growing points per unit length of RNA, and the proportion of copies that are complete and accurate. Each of these capacities can in turn be derived from the free energies of base-pair formation, as discussed by Eigen (1971).

There is more to the fitness of the molecule than the physical chemistry of complementary base pairing, because it can fold into three-dimensional shapes, and these shapes affect each of the capacities represented by $b_i$, $d_i$, and $R_i$. An open shape would be most conducive to high rates of replication and high $b_i$, since a closed structure would inhibit the establishment and growth of the replication point. At some point, folding might help to displace the newly synthesized daughter strand and permit a second and third replication, so long as it occurred after synthesis was complete. So a cycling of conditions would likely be needed for continued template-mediated replication without the aid of enzymes.

The resource utilization parameter $R_i$ measures the effectiveness of the molecule in obtaining and processing resources. Three issues are important. The first is the likelihood of forming a correct hydrogen-bonded association when a free mononucleotide interacts with its complement in the template RNA. The second is the probability that the new nucleotide will link up with the growing daughter strand, and the third is the stability of the newly linked nucleotide before the next nucleotide arrives and continues the process. Again, these processes will be determined by the free energies of base pairing and the shape and folding of the molecule in its environment. The death rate of the molecule, $d_i$, is determined primarily by hydrolysis. Again, the free energies of base pairing (Eigen 1971) and the shape of the molecule will be paramount—a more compact shape being the more stable and protecting the molecule from death.

Without making proteins, the molecular replicator would likely encounter constraints in simultaneously maximizing birth and survival. A more open structure facilitates birth but risks death in a hydrolytic environment. Consequently, a fluctuating environment is probably most conducive to the origin of life, if indeed life began as envisioned here by a template molecule replicating via complementary base pairing (Kuhn 1972). For example, if the temperature changes with the daily cycle, or if hydration and dehydration alternate with the tides, the conflicting needs of birth and death might be satisfied at different times.

Regardless of the practical details, I hope it is clear that a structural analysis of the design features of the template molecule is possible, and that such an analysis serves to deconstruct the heritable capacities in table 2-1 and equation 2-2 into more specific features of the molecule. We can thus understand the attributes of the molecule that underlie its Fisherian fitness by studying the heritable capacities, $b_i$, $d_i$, and $R_i$. By virtue of their increased rate of growth, those replicators that optimally balanced the conflicting demands of birth, survival, and resource utilization would come to predominate in the population. With the emergence of self-replication, life history evolution began (Michod 1983b).

## LIFE HISTORY EVOLUTION

The conditions for selection among a community of replicators may be derived from equation 2-2, so long as all replicators follow a similar dynamic. I assume that the level of available resources, $R$, is a monotonically decreasing function of the total density of replicators in the community, $\sum X_i$. Define the carrying capacity $\hat{K}_i$ of each replicator $i$ as the density the replicator would attain if it were alone in the environment at equilibrium. A simple but general analysis of equation 2-2 shows that the condition for a new replicator of type $j$ to invade from rarity a community dominated by type $l$ is

$$\hat{K}_j > \hat{K}_l. \tag{2-3}$$

I interpret $X_i$ as the number of mononucleotide building blocks tied up in molecules of type $i$. Thus I may write the available resources as the total resources if there were no replicators, $R_T$, minus the resources tied up in replicators, to obtain available resources as $R = R_T - \sum X_i$. This way of parameterizing resources gives the carrying capacity of type $i$ as

$$\hat{K}_i = R_T - \frac{d_i R_i}{b_i - d_i}. \tag{2-4}$$

Upon substituting equation 2-4 into equation 2-3, I obtain the more familiar condition for selection:

$$\frac{b_j - d_j}{d_j R_j} > \frac{b_l - d_l}{d_l R_l}. \tag{2-5}$$

Selection among molecules using template-mediated replication is a kind of $K$ selection (MacArthur 1972), where it is understood that $K$ is a function of all the heritable capacities as indicated in equation 2-4.

## SURVIVAL OF THE FITTEST

Having defined the components of fitness for the molecular replicator, $b$, $d$, and $R$, can we combine them into an overall measure of fitness? What would we want to do with such a measure? Predict the outcome of natural selection, of course.

The condition given in equation 2-5 for selection of one kind of replicator over another involves only the heritable capacities of the replicators concerned. Consequently, the particular combination of heritable capacities given in equation 2-5, $(b_i - d_i)/d_i R_i$, can be considered to be an overall measure of adaptedness of the different kinds of molecules. In other words, whether one type will be selected depends only on its heritable capacities in the manner specified in equation 2-5. The fittest type, in the sense of having the largest $(b_i - d_i)/d_i R_i$, survives the best over evolutionary times.

This is exactly what I think Spencer and Darwin meant by the phrase "survival of the fittest." In this phrase, "survival" means evolutionary success, and "fittest" means best designed.[1] Natural selection boils down to evolutionary success of the best designed, or at least this is what "survival of the fittest" thinking means. I think it was probably for this reason that Eigen and his colleagues termed this kind of natural selection "Darwinian."

25

I also think this is what Fisher was searching for when he tried to express fitness, in the sense of the per capita rate of increase, in terms of the birth and death capacities (recall the discussion above of equations 1-1 and 1-2).

The result that the heritable capacities by themselves predict the outcome of selection is important. Selection depends only on individual capacities, and a measure of adaptedness of the individual molecule is possible in this case, namely, $(b_i - d_i)/d_i R_i$ in the specific case considered here (equation 2-5). Unfortunately, this deconstruction works only for simple cases of Malthusian growth, in which birth is a linear function of density $X_i$, $b_i X_i$. The form of the replicator dynamic critically affects the outcome of selection.

We used the term "Malthusian" to refer to this kind of replicator dynamic (Bernstein et al. 1984; Michod 1983b), even though Eigen had previously used the term "Darwinian" to refer to the same kind of dynamic. As just mentioned, Eigen's terminology makes sense because the character of selection is aptly described by Darwin's principle of survival of the fittest. However, this terminology has the unfortunate consequence of suggesting that other forms of selection dynamics that are not linear in density (such as sexual reproduction) are not "Darwinian," even though they obey Darwin's principles of variation, heritability, and competition. While I believe that Eigen's attempt to classify selection according to the dynamics involved greatly expands our understanding of the process of natural selection, I find his terminology here misleading. For this reason, I use the term "Malthusian" to refer to selection dynamics involving birth processes that are linear in density. In later chapters, I consider the evolution of protein-mediated replication and the evolution of sexual reproduction. In these cases, the birth process becomes nonlinear in $X_i$ and this dramatically changes the character of selection. But for now, we have a comforting result—the heritable capacities of the unit of selection (in this case a replicating molecule) predict the outcome of natural selection. Darwin was right: in this case, the fittest (best designed, or most adapted in the sense of equation 2-5) do survive the best over evolutionary time, and it is possible to provide a measure of overall adaptedness (equation 2-5) that combines adaptive features of the replicator in a way that predicts evolutionary success.

## SURVIVAL OF ANYBODY

The Malthusian form of birth assumed in equation 2-2 leads to exponential increase in the absence of resource regulation. Another approach to template replication assumes a subexponentially growing population of nucleic acid templates (that is, the population growth is slower than exponential), in which the replication process, instead of being linear in density, $X_i$, grows according to the square root of density, $b_i X_i^{1/2}$ (Szathmáry 1991; Szathmáry and Maynard Smith 1997). One reason for subexponential growth might be that birth is inhibited by the complementary binding of template and copy. There are merits to both approaches (assuming either exponential or subexponential growth) but this does not concern me here. More interesting are the implications of this subexponential growth law for selection (Michod 1991; Szathmáry 1991). When replicators grow according to the square root of density, instead of linearly in density, any new mutant type will increase when rare regardless of its heritable capacity at birth, $b_i$.

In this case there is no measure of overall adaptedness, since evolutionary success does not depend upon the heritable capacities of the replicator. By analogy with Darwin's "survival of the fittest" phrase, this outcome may be described as the "survival of anybody." "Survival of anbody" may suggest biological nonsense, but this outcome is potentially important because the underlying cause is a cost of commonness (Michod 1991). The cost of commonness can be a diversity-producing force in communities, for example, when parasites fixate on common host types.

The point here is that the dynamics of selection depend on the dynamics of population growth, and this critically affects the outcome of selection. Survival of the fittest is just one of many possible outcomes. There is more to selection and evolutionary success than increase of the better-designed phenotypes.

## OVERVIEW OF THE ORIGIN OF FITNESS

Order emerged from the chaos when molecules became capable of self-replication. Once replication began, fitness relations could further direct this transition into the living realm. Although it is possible to discuss the fitness of any physical object in terms of its persistence over time, biological fitness requires a life cycle involving replication and death, that is, fitness requires a life history.

The way in which birth depends upon density affects the form of the Darwinian dynamic, and different kinds of dynamics give rise to different outcomes and characterizations of the process of natural selection. As modeled here, template-mediated self-replication is a Malthusian growth process, in which birth is assumed to depend upon replicator density in a linear fashion, although other forms of birth, such as subexponential growth, are

27

equally likely (Szathmáry 1991; Szathmáry and Maynard Smith 1997). Survival of the fittest describes the outcome of natural selection only when birth is linear in density, not otherwise.

The fact that the condition for selection for Malthusian replicators depends only on the heritable capacities as shown in equation 2-5 suggests $(b_i - d_i)/d_i R_i$ as a measure of overall adaptedness of replicator type $i$. It combines the heritable capacities at birth, death, and resource utilization in the manner specified in equation 2-5 into an overall measure of adaptedness that predicts evolutionary success. In this way, we see how overall adaptedness can be used to predict the outcome of natural selection by combining heritable capacities into a predictor of evolutionary success. This is the essence of the adaptationist paradigm as encapsulated in the phrase "survival of the fittest," where "survival" refers to survival over evolutionary time or evolutionary success, and "fitness" refers to adaptedness or good design. I think this is what Darwin intended when he adopted the phrase from Spencer to describe what he expected as the outcome of natural selection. A design analysis of the structure of a molecule replicator shows what features likely determine fitness. We can thus understand the design of the molecule in a way that is dynamically sufficient, by studying the heritable capacities at birth, $b_i$, at survival, $1 - d_i$, and at utilizing resources $R_i$.

This property of predicting evolutionary success from adaptive features holds only for Malthusian replicators, in which birth is linearly related to density. As already mentioned in the case of subexponential growth, any new type can increase when rare, regardless of its heritable capacities (survival of anybody, or cost of commonness). In the next chapter, I consider protein-catalyzed replication; in this case, as with sexual reproduction, birth depends upon density in a quadratic nonlinear fashion (replication proportional to $X_i^2$). We discover that the outcome of natural selection is dramatically changed again, but now there is a cost of rarity leading to an outcome aptly described as "survival of the first." No type can increase when rare regardless of its heritable capacities (Bernstein et al. 1985c; Hopf and Hopf 1985; Michod 1983b). The main point of all these examples is that when birth is a nonlinear function of density, the heritable features of a unit of selection are no longer sufficient to predict the outcome of natural selection. Survival of the fittest is false, and there is no useful measure of overall adaptedness.

# The First Individuals

## ORIGIN OF GENE NETWORKS

In this chapter I consider the further evolution of molecular replicators, first as they formed groups of cooperating genes (the first genome) and then as they formed the first cell. Let us return to the situation of short-length oligonucleotide sequences replicating by template polymerization. The continued evolution of these replicators would lead to sequences that balanced the conflicting demands of replication and survival. However, there were limits to what any one sequence could attain by itself. Without separate enzymes to increase the fidelity of replication, the length of the sequence was severely constrained. Even before the evolution of the genetic code and a way of making proteins, separate ssRNA molecules could possibly aid each other's replication by providing catalytic surfaces facilitating nucleotide pairing and linkage. The transition from molecular replicators to networks of cooperating genes was the first major transition in evolution. Eigen and Schuster have developed a theory to explain how such cooperating groups of genes, termed "hypercycles," could emerge in communities of molecular replicators (Eigen 1992; Eigen et al. 1981; Eigen and Schuster 1979, 1978a,b, 1977). Hypercycles are groups of interacting genes and serve as a model for the organization of the primitive genome. Here I am concerned with how cooperative hypercycles start in mixed populations with Malthusian replicators, which must have been the situation when cooperation first arose in populations of replicating molecules.

My primary concern in this chapter is to understand the principles that guided the transition from Malthusian replicators (single genes) to groups of genes (hypercycles) as a model for the creation of new units of selection and fitness in evolution. The transition from genes to cooperative gene networks shares fundamental elements of later evolutionary transitions (hypercycle to cell, cell to organism, organism to society), so let us consider the general properties of groups of cooperating genes carefully.

There are two levels of selection and two levels of fitness effects, the gene and the gene group. I am interested in the transition between these levels, and, specifically, how fitness emerges at a new organizational level, the group of genes, while reducing the selective demands of the elements in the group, the individual genes. Cooperation accomplishes this because it trades fitness effects from the genes (the cost of cooperation) to the group (the benefit of cooperation). Because cooperation and its antithesis, defection, trade fitness between levels, they are fundamental to the study of evolutionary transitions (table 3-1). Viewed in the present context, cooperation (and altruism) and defection are not just special problems in the study of animal behavior, but rather a central issue in transitions to increased complexity in evolution.

Genes interact either directly or through the effects of the proteins they make. These effects add to or detract from the gene's capacity to replicate and survive (table 3-1). Before the creation of chromosomes or a cell, these interactive effects are frequency and density dependent, because the likelihood of interaction between two gene sequences, or a gene sequence and its neighbor's protein, depends upon the product of the densities of each. Frequency-dependent selection will be considered in general terms in the next chapter. Suffice it to say here that there is always an advantage to cheating within the group, and this can frustrate progress to higher levels of organization. Because the benefits of cooperation are frequency dependent while the costs are not, overcoming the limitations of frequency-dependent selection is a central problem in the creation of new levels of organization.

Cooperative gene networks can have any number of gene members. In figure 3-1, one- and two-membered hypercycles are shown along with a Malthusian replicator replicating by template-mediated growth. As already discussed in the last chapter, the rate of template-mediated replication is proportional to $b_i X_i$. However, the rate of enzyme-mediated replication is taken to be proportional to $B_i X_i^2$ (for a one-membered hypercycle) or $B_2 X_1 X_2$ (for replicator 1 in a two-membered hypercycle), because the replicator must encounter the enzyme, and there is no cell membrane at this early

| | Level of Selection | |
| Gene Behavior | Gene | Group (hypercycle) |
| --- | --- | --- |
| Defection | (+) Replicate faster or survive better | (−) Less functional |
| Cooperation | (−) Replicate slowly or survive worse | (+) More functional |

*Table 3-1*

Effect of cooperation on fitness at gene and hypercycle level. The +/− symbol means positive or negative effects on fitness at the gene or group level.

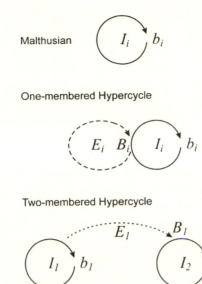

Malthusian

One-membered Hypercycle

Two-membered Hypercycle

*Figure 3-1*

Birth in hypercycles and Malthusian replicators. The variables $I_i$ and $E_i$ stand for informational molecule (the gene) and enzyme, respectively, of replicator type $i$. $B_i$ is the beneficial effect of the enzymes produced by type $i$ on the rate of replication of another replicator. The rate of template-mediated replication is $b_i$.

stage to keep the enzyme close to the replicator (see figure 3-1). Because of this nonlinearity, hypercycles are not Malthusian replicators and they may show hyperbolic growth. Spatial structure may be present, but for the time being I am considering a well-mixed system.

These early networks of genes were a poorly defined lot. However, this lack of individuality may have had at least one advantage (Michod 1995; Bernstein et al. 1984). It would be very easy for genes to become damaged, exposed as they were without a cell membrane to the ultraviolet radiation and chemical reactions in the primordial soup. Breaks, loss of nucleotides, and loss of methyl groups are just a few of the kinds of damages that occur to DNA in modern organisms and likely threatened the existence of these early gene networks. However, mixed up as we imagine these early genes to be, without any cell membrane to trap the errors, damaged genes would just fail to replicate and their undamaged brethren could make an extra copy to fill in.

31

Sex (mixing and repair of genes) likely came easily to these early replicators. There was much promiscuity and little individuality at this early stage.

## COOPERATION AND CONFLICT

Two issues are central to the creation of a new unit of selection—promoting cooperation among the lower-level units in the functioning of the group, while at the same time finding ways of mitigating the inherent tendency of the lower-level units to compete with one another through their frequency-dependent effects. Cooperation represents the benefit of group living: groups of molecules can behave in new and useful ways; two molecules can do more than one. There is another consequence of cooperation: it is usually costly to the fitness of the individuals involved. Cooperation is a critical factor in the emergence of new units of selection because it trades fitness at the lower level (its costs) for increased fitness at the group level (its benefits). In this way, cooperation can create new levels of fitness (table 3-1).

Molecular replicators interact through protein products and/or catalytic surfaces. Besides being the means for cooperative interactions and higher-level functions, the evolution of proteins allowed the genotype and phenotype to diverge, with each specializing in a distinct purpose: information storage and transfer for the genotype and instrumentality and function for the phenotype. Prior to the evolution of the protein, the phenotypic functions of the molecular replicator were directly tied to its physical structure. With the decoupling of phenotype from genotype a greater flexibility and diversity of instruments of purpose and design became possible.[1]

The benefit of cooperation among replicators in the hypercycle is represented by the parameter $B$ (benefit), which measures the capacity of the enzyme (or catalytic surface of the ssRNA) to aid the replication rate of other gene sequences (see figure 3-1 and equations 3-4, 3-5, 3-7, and 3-11 below). Although group living has benefits for those involved, it provides the opportunity for conflict and cheating. How might conflict destroy the hypercycle? Selfish replicators that maintain good target functions for the enzymes of others but refuse to make their own enzymes are but one example. Unbridled competition among molecules would lead to cheaters who reap the benefits of their neighbors' costly efforts, while producing nothing useful for the group themselves. This would threaten the viability of cooperative molecules, leading to a loss of cooperation and the benefits of group living. The tragedy of the commons existed right on life's doorstep.

Garrett Hardin (1968) identified the tragedy of the commons as the central problem of group-beneficial behavior. When individuals stand to gain more by behaving selfishly than their selfish behavior costs each member of the group, there is a temptation (read immediate selective advantage) to cheat. In cooperative groups such as networks of cooperating genes, there is always a temptation for individuals to defect. What keeps defection from arising and taking over the group, ruining the network of cooperation upon which its very advantage depends? Human societies have laws and police forces to reduce this temptation. What happens during evolutionary transitions? How can group-beneficial behavior ever evolve? In short, the group must become an individual, but how does this happen?

Natural selection cannot usually be counted on to lead to fitness-enhancing behaviors such as cooperation.[2] It is too short-sighted for that. In well-mixed populations, natural selection leads to defection and selfishness, even though this results in decreased fitness for all concerned. Only under certain conditions can cooperation gain a foothold in the population and flourish. What are these conditions and how do they arise? This question is considered in a general context in the next chapter, but let us now consider simple populations of replicating genes.

## SURVIVAL OF THE FIRST

To investigate the problem of the origin of cooperative networks of genes, or hypercycles, I use Eigen and Schuster's form of resource regulation by defining in equation 3-1 the per capita production of the community of replicators, $\Psi$ (with the understanding that the unlimited rate of growth is to be used for $\dot{X}_i$ in equation 3-1):

$$\Psi = \frac{\sum_i \dot{X}_i}{\sum_i X_i} \tag{3-1}$$

A molecule replicating by template base pairing is described by the Darwinian dynamic in equation 3-2 for this model of resource regulation:

$$\dot{X}_i = X_i(b_i - d_i - \psi) \tag{3-2}$$

This representation of resource competition is simpler than equation 2-2 since there is no replicator-specific resource utilization parameter, just a kind of mass action effect embodied in $\psi$ so as to keep the total level of replicators constant. Eigen's replicator dynamic is still Malthusian since birth is linear in replicator density. So the condition for selection for equation 3-2 (analogous to that given in equation 2-5) is still a function only of the heritable capacities (now $b_i$ and $d_i$) and corresponds to survival of the fittest. The condition for selection is given by

$$b_j - d_j > b_l - d_l. \tag{3-3}$$

Now let us consider evolution among hypercycles. We find that the non-linearities inherent in hypercyclic replication give a different character to natural selection, different from the survival of the fittest and survival of anybody discussed in the last chapter. Consider competition between two one-membered hypercycles that replicate only with the aid of enzymes (no template-mediated replication, $b_i = 0$) as described by the following pair of differential equations:

$$\dot{X}_1 = (B_1 X_1 - \psi) X_1,$$

$$\dot{X}_2 = (B_2 X_2 - \psi) X_2. \tag{3-4}$$

Equation 3-4 satisfies Darwin's conditions for natural selection (there is variation, heritability, and resource limitation), but analysis of this competition shows that no mutant type can increase when rare—no matter what its individual capacities are. The equilibrium point at which one type (say, replicator 1) predominates is always stable to the invasion by replicator 2 (assuming $B_1 > 0$). This is because when replicator 1 is at equilibrium, the density of type 2 is vanishingly small, and there is an inherent cost of rarity because reproduction requires an interaction between like types. This outcome ștems from a cost of rarity which is one of the problems inherent to frequency-dependent selection on cooperation.

In the last chapter (and in equation 3-3), we found that when population growth is exponential new types may evolve according to their individual capacities, and survival of the fittest accurately describes the outcome of selection. When population growth is subexponential, any new type can increase regardless of its individual capacities. This is a diversity-producing mechanism in evolution and may be termed survival of anybody. Now we find that when population growth is greater than exponential, as it is with protein-mediated replication or when there is sex, new mutant replicators cannot invade—no matter what their heritable capacities. Their enzyme may be better designed; nevertheless, such improved replicators cannot increase from rarity, because these benefits are proportional to the square of density. Common types persist over time by virtue of their abundance, not their design. Eigen described this situation as "once and forever evolution," since new types have difficulty invading (1971). By analogy with Darwin's phrase "survival of the fittest," my colleagues and I have characterized this evolution as "survival of the first" (Bernstein et al. 1984; Michod 1983b). Again, we find that the dynamics of natural selection involves much more than the increase of the better-designed phenotypes.

## EVOLUTIONARY TRANSITIONS ARE INHERENTLY NONLINEAR

The nonlinearity inherent in hypercycles is not only a property of gene networks at this early stage in life; rather, it is a general feature of evolutionary transitions. New higher-level units gain their properties by virtue of the *interactions* among lower-level units. Before the evolution of a structure to "house" the new higher-level unit (and this must come later), interactions among lower-level units are density and frequency dependent, and, therefore, there will be problems with rarity and advantages to commonness.

Eigen and coworkers have (correctly) attached great importance to these nonlinearities and their consequences for selection (Eigen and Schuster 1979; Eigen 1971; Eigen and Schuster 1977, 1978a,b). They emphasize the intrinsic advantages of hypercycles in competition with Malthusian replicators in that "selection of a hypercycle is a 'once-forever' decision" (Eigen and Schuster 1978a, p. 41). However, the transition from template-mediated replication to hypercyclic networks occurred gradually in mixed communities of hypercycles and Malthusian replicators. Furthermore, template-mediated replication was presumably not discarded immediately with the evolution of proteins. All this is to say that the appropriate framework for understanding the origin of cooperative groups of genes is mixed systems of replicators possessing both Malthusian and hypercyclic components. I now consider this situation.

## ORIGIN OF HYPERCYCLES

Since template-mediated (nonenzymatic) replication arose before enzyme-mediated replication, I study the competition of hypercycles and Malthusian replicators. Consider competition between a one-membered hyper-

35

cycle (with both template and enzyme components to replication) and a Malthusian replicator as embodied in equation 3-5:

$$\dot{X}_1 = (r_1 - \Psi)X_1,$$

$$\dot{X}_2 = (r_2 + BX_2 - \Psi)X_2,$$

$$\Psi = \frac{r_2 X_2 + BX_2^2 + X_1 r_1}{X_T}, \tag{3-5}$$

with $X_1 = X_T - X_2$ and $r_i = b_i - d_i$.

I assume here that the enzyme is specific for the replicator that produced it (type 2). This assumption clearly works to the advantage of the hypercycle, but it may be unrealistic if the first enzyme or enzymelike region of the ssRNA was a more general-purpose surface that could be utilized by any replicator. I relax this assumption later in the chapter when the consequences of spatial structure are investigated. There is a significant difference (with regard to the issue of density-dependence) between making an enzyme as a separate product and utilizing a region of the ssRNA molecule in an enzymelike manner. A replicator that is able to utilize one region of its ssRNA molecule as a catalytic surface presumably does not have to encounter this region in a mass action (density dependent) manner. This effect would be included as an increase in the Malthusian birth terms represented by $b_i X_i$. If, however, this same region were to be used by another replicator, the interaction would involve mass action dynamics and be represented in the hypercyclic term $B_i X_i X_j$. In any event the situation represented in equation 3-5 is that of a one-membered hypercycle (with a Malthusian component, $r_2$) in competition with a Malthusian replicator.

When the hypercycle (type 2) is rare, its ability to invade a community dominated by the Malthusian replicator (type 1) is determined by $r_1$ and $r_2$. If there is any cost to producing the protein, the hypercycle cannot invade since $b_2 < b_1$. What about when the hypercycle type 2 is common and type 1 is rare? In this case the hypercycle is stable if

$$BX_T + r_2 > r_1. \tag{3-6}$$

The Malthusian replicator will increase if equation 3-6 is not satisfied. This may be difficult, especially if the carrying capacity is large and the rates of enzyme-mediated replication are high (implying that $BX_T$ is large). Enzyme-mediated reactions are typically many orders of magnitude greater than their nonenzymatic counterparts; however, with early enzymes this may not be true. On the other hand, the costs of enzyme production (paid in the $b_2$ term) may be large if the early attempts at making enzymes were time-consuming and inefficient. This would help the evolutionary prospects of the Malthusian replicator.

In the case of competition between a Malthusian replicator and a two-membered hypercycle, the corresponding condition to equation 3-6 indicates that it is twice as easy for the Malthusian replicator to increase, all other factors being equal (Michod 1983b). Two-membered hypercycles must share available resources among both components, leading to a lower density of each component (by about a factor of one-half), and since the benefits of the hypercyclic interactions are proportional to density this decreases their stability by about the same factor.

We have discovered that evolution of a hypercycle is not a "once-forever" decision, as Eigen envisioned in communities with mixed Malthusian and hypercyclic components. In addition, when hypercycles are rare, their Malthusian birth terms dominate and any cost to enzyme production would detract from their evolutionary prospects, since the costs would be paid in the Malthusian term. How then can a hypercycle increase and lead to co-operative networks of genes? There are two ways to overcome the afore-mentioned barriers to the origin of cooperative gene groups and the costs of cooperation in this system. The first is based on Eigen's concept of a qua-sispecies distribution; the second is based on spatial structure and kin se-lection.


QUASISPECIES

Because of the high error rates inherent in primitive replicating systems, mutation was likely a dominant force in the evolution of early replicators. Eigen recognized this fact by assuming that a dominant or "master" se-quence would produce (via mutation) a cometlike tail of mutant types. The complete ensemble of master sequence and mutant types is termed the *qua-sispecies* distribution. How might the quasispecies distribution help us un-derstand the origin of hypercycles? I assume that a Malthusian replicator, the master sequence, produces another hypercyclic replicator through re-current mutation. It is the mutant that produces the enzyme, so the master sequence does not pay the costs of the enzyme. The hypercycle may avoid the costs of rarity since it need not originate from rarity. It may hitchhike to significant frequency by virtue of the attributes of the master sequence. Once it reaches a significant density, its own attributes may direct its con-tinued evolution. Can such a scenario work, at least in theory?

Consider a very simple case of master sequence in density $X_1$ with one mutant type in density $X_2$.[3] The mutant type 2 is assumed to be a one-membered hypercycle as diagrammed in figure 3-1. The birth term in the $\dot{X}_2$ equation due to the protein will be a function of protein density and $X_2$. I assume that the protein concentration follows $X_2$ adiabatically, so that the overall growth rate due to the protein is $BX_2^2$. The mutation rate is $\mu$. The dynamical equations then read

$$\dot{X}_1 = X_1(r_1 - \mu - \Psi),$$

$$\dot{X}_2 = X_2(r_2 + BX_2 - \Psi) + \mu X_1,$$

$$\Psi = \frac{r_2 X_2 + BX_2^2 + r_1 X_1}{X_T}. \tag{3-7}$$

In finding the steady states and equilibrium properties it is convenient to define the variable $z_2 = X_2/X_T$ and to let $X_1 = X_T - X_2$. The steady states are then given by

$$(z_2 - 1)(BX_T z_2^2 - (r_1 - r_2)z_2 + \mu) = 0. \tag{3-8}$$

Fixation of the hypercycle, $z_2 = 1$, is locally stable if equation 3-9 holds:

$$BX_T + r_2 > r_1 - \mu. \tag{3-9}$$

The left-hand side of equation 3-9 is the growth rate of the mutant type 2 when it is at carrying capacity, while the right-hand side is the growth rate of the master sequence when rare. Equation 3-9 also gives the condition for instability of the complete quasispecies (both master sequence and mutant hypercycle). If there is no mutation, equation 3-9 simplifies to equation 3-6.

The results of the quasispecies origin of a hypercycle are shown in figure 3-2, in which the equilibrium frequency of the hypercycle and its stability are plotted as a function of the maximum benefit of enzyme-catalyzed replication. At first the quasispecies is stable and the hypercycle is not stable, but as the benefit of enzyme-catalyzed replication increases there is an intermediate region of parameter values for which fixation of the hypercycle and the quasispecies are both locally stable. This region is termed "bistable" and is described in figure 3-2. Because of fluctuations in nature, the system may alternate between the quasispecies and the hypercycle. As the benefit of the enzyme increases further, the hypercycle replaces the quasispecies. Consequently, it is possible for the hypercycle to originate as a member of a quasispecies. I now consider the role of population structure in the origin of hypercycles.

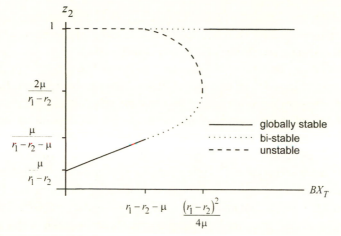

*Figure 3-2*

Quasispecies origin of the hypercycle. Equilibrium frequency of the hypercycle (plotted on the ordinate axis) as a function of the maximum benefit of enzyme-aided replication (the abscissa).

## POPULATION STRUCTURE

What factors help control the spread of cheaters and interlopers of the hypercycle? Population structure and genetic kinship are two factors of almost universal generality. The origin of life field has often focused on the origin of cell-like structures such as coacervates (Oparin 1968, 1965) or proteinoid microspheres (Fox and Harada 1960). These structures provide for the localization of protein products and help contain the spread of selfish molecules that use the proteins made by others but do not contribute their own protein to the well-being of the group. The perspective taken here is that replicating molecules likely predated the emergence of cell-like structures; nevertheless, the emergence of the protocell remains an evolutionary milestone. Can one discuss a continuous evolution leading to the cell, or did it have to emerge in one discontinuous leap?

The primordial "soup" envisioned by early workers in the field (Haldane 1929; Oparin 1938) was not a blended homogeneous broth but rather a chunky mix of discontinuous materials and structures. Rock crevices and suspended clay particles could provide local habitats for the hypercycles (Van Holde 1980). One line of thinking argues that life arose in the atmosphere in suspended water droplets within which molecules could be distributed by the wind (Towe 1981; Woese 1980). "In this atmosphere all stages in evolution are basically 'cellular.' The droplet phase serves as a natural definition of the protocell" (Woese 1980, p. 68).

Spatial effects can radically alter the outcome of selection, especially the likelihood of invasion of new types. Several studies have shown that invasion can be very sensitive to local conditions when space is explicitly included. Spatial effects can facilitate the evolution of cooperation, a result at odds with common wisdom (Ferriere and Michod 1996, 1995). Another example of selection reversal due to spatial effects was provided in a host-parasite model (Hassell et al. 1991) that was subsequently studied in an evolutionary context using cellular automata (Boerlijst and Hogeweg 1993). In this model, whether mutant parasites are able to invade is determined by the expansion of self-generated spatial patterns (spiral waves). A less efficient parasite can nonetheless wipe out the more virulent wild parasite type if two conditions are met: (i) the parasite's population initially develops as a spiral wave whose center is located far from the edges of the metapopulation and (ii) the mutants appear at the very center of the spiral. Similar results help hypercycles resist the invasion of parasitic mutants (Boerlijst and Hogeweg 1991).

When space is explicitly included in models of the evolution of interactions, the outcome of selection can no longer be understood simply at the level of the individual fitnesses of phenotypes or genotypes involved in the competition. To predict the success of an invasion attempt, one needs to consider the local dynamics of the population (where the spiral centers are located, how fast the spirals rotate, etc.). This is to say that the environment is involved in the selection process, not only in the sense of an external agent to which the population adapts but as an integral part of the selection process. The environment is the population. (The implications of this fact for a general understanding of "fitness" in evolution will be taken up in the last two chapters.)

## Kin Selection in Evolutionary Transitions

Population structure provides conditions conducive to maintaining genetic relatedness among gene sequences. Kin selection can help reduce the effectiveness of defection and cheating as viable strategies in groups of gene replicators. I now apply D. S. Wilson's group selection model (Wilson 1975, 1980) to the problem of hypercycle evolution, following an earlier paper of mine (Michod 1983b). This approach has recently been extended (Maynard

Smith and Szathmáry 1995; Szathmáry and Maynard Smith 1997). The densities $X_1$ and $X_2$ now represent the global densities of two replicators that are distributed with probability density $P(y)$ into local habitats each of size $N$. The probability density function $P(y)$ may be interpreted as the density of habitats with a particular number $y$ of type 2 replicators ($y = 1, 2, \ldots,$ $N$) with $N - y$ being the number of type 1 replicators. Replicators of type 2 are assumed to make a primitive protein that benefits the replication of both replicators, in particular, type 1 replicators, who do not make proteins. The primitive protein catalyst is produced at a cost $C$ to replicator type 2; its local density follows the density of type 2 adiabatically and so is proportional to $y$.

Even though both replicators may use the protein, the "experienced densities" of the protein, $e_i$, are different for the two replicators. Equation 3-10 gives $P(y/i)$ the conditional local density of type 2 for a replicator of type $i$ ($i = 1,2$) and the experienced densities $e_i = \sum_y yP(y/i)$:

$$P(y/1) = \frac{(N-y)P(y)}{\sum_y (N-y)P(y)} \Rightarrow e_1 = \bar{y} - \frac{\sigma_y^2}{N - \bar{y}}$$

$$P(y/2) = \frac{yP(y)}{\sum_y yP(y)} \Rightarrow e_2 = \bar{y} + \frac{\sigma_y^2}{\bar{y}} \tag{3-10}$$

I use the experienced densities of the protein given in 3-10 in the replicator dynamics to obtain the following system of equations describing evolution of hypercycles in a structured population.

$$\dot{X}_1 = X_1(r_1 + Be_1 - \Psi)$$
$$\dot{X}_2 = X_2(r_2 - C + Be_2 - \Psi)$$
$$\Psi = \frac{X_1(r_1 + Be_1) + X_2(r_2 - C + Be_2)}{X_T} \tag{3-11}$$

Analysis of equation 3-11 gives the following results (Michod 1983b). The hypercycle will increase if

$$\frac{C}{BN} < F_{ST}, \tag{3-12}$$

where $F_{ST} = \sigma_y^2/\bar{y}(1 - \bar{y})$ is defined by Wright (1951) to be the ratio of the observed variance to the maximal possible variance if all localities were fixed for one or the other type of replicator. Equation 3-12 is analogous to

Hamilton's rule (Hamilton 1963, 1964a,b) commonly derived in models of kin selection (for a review see Michod 1982). Wright's $F_{ST}$ statistic measures the degree to which the community of primordial replicators is structured—it is the correlation between two replicators picked at random from a locality relative to those picked at random from the global population.

Population structure is important because it leads to nonrandom associations, resulting in heightened genetic relatedness between interacting types. This is conducive to cooperation. Consequently, evolution in spatially structured populations involves kin selection, and the underlying genetic relatedness serves to facilitate the evolution of cooperating hypercycles even when pitted against noncooperating types such as replicator type 1 above.

Kin selection plays an important role in evolutionary transitions. New evolutionary units are "individuals" in the sense of being spatially and temporally localized. Consequently, we may expect spatial structure and its concomitants genetic relatedness and kin selection to play a critical role in the origin of new units of selection. During the emergence of a new unit, population structure (discussed above), local diffusion in space (Ferriere and Michod 1996, 1995), and self-structuring in space (Boerlijst and Hogeweg 1991) may facilitate the trend toward a higher level of organization, culminating in an adaptation that legitimizes the new unit once and for all. Examples of such adaptations include the cell membrane in the case of the transition from genes to groups of cooperating genes, or, as discussed in chapter 6, the germ line or self-policing functions, in the case of the transition from cells to groups of cooperating cells, that is, multicellular organisms.

## CONFLICT MEDIATION THROUGH INDIVIDUALITY

There was an advantage to localizing products for the use of the replicators that made them. Localization of proteins would make them more efficient and effective; more importantly, it would help protect the cooperative unit from cheaters and interlopers. Population structure, local diffusion in space, and self-structuring in space may have facilitated the evolution of cooperation and the transition to more cooperative networks of genes, but ultimately the hypercycle would solidify its private existence by making a protective cellular membrane and becoming the first new individual, the cell.

The cell was a wonderful invention; it protected the genes from the damaging effects of the environment and allowed resources and proteins to be kept close at hand, instead of diffusing away to be used by others. By encapsulating gene networks into a cell-like structure, proteins and other products would only be available to those neighbors sharing the same cell. If one of these neighbors turned selfish (say, by using its neighbors' proteins but not taking time to make it own), all the genes in the same cell would be threatened, including the selfish gene. By putting everybody in the same boat, so to speak, everybody's self-interest becomes more closely aligned with the interests of the group. In this way, population structure serves the emergence of higher-level groups and units of selection.

Increasing complexity requires more inclusive individuals that regulate the selfish tendencies of their components—genes in the case of gene networks, component cells in the case of multicellular organisms. How is individuality created at a new level (cell or organism) so that the lower-level units may be regulated, especially when there is no external agent that sits outside the system to be in control?

The cell is an example of a device that reduces conflict because it more closely aligns the interests of the members (genes in the present case) with the interests of the group (hypercycle). For all the genes to replicate, the cell has to reproduce. So each gene has an interest in promoting the replication and well-being of the cell. To understand the evolution of individuality and new levels of organization we need to identify those mechanisms and structures that serve to align the interests of the component parts with the interests of the group. During the origin of each new kind of individual, conflict mediation is a necessary step, otherwise new adaptations at the new level cannot evolve, for there is no clearly recognizable (by selection) unit, no individuality. The evolution of conflict mediation is necessary for adaptation at the new level.

## FURTHER EVOLUTION OF THE CELL

With the evolution of the cell, new problems emerged, of which two were especially critical. First, genetic errors once repaired by the free exchange among communities of genes became trapped inside the cell. Second, a way had to be found of segregating a complete complement of genes into offspring cells. Both of these problems stem from the loss of gene function, either through genetic damage and mutation or through actual loss of genes during the early attempts to package the complete hypercycle in daughter cells. One solution to the segregation problem, chromosomes, further aggravated the problem of genetic damage, as discussed below.

Elsewhere I have discussed the theory that sex among cells (mating and

43

recombinational repair, followed by splitting) serves to repair damaged genes while masking the effects of deleterious mutations (Michod 1995). Sexual exchange, which had come easy to the free-living hypercycle, had to be reinvented once the cell evolved (see Michod 1995, pp. 16–32). Sex between cells involved mating (fusion) of two cells, followed by mutual repair (recombination) and then splitting into daughter cells again. Diploidy (staying fused and carrying a set of genes along) was another possible strategy to cope with genetic error (Michod 1998; Long and Michod 1995; Michod and Long 1995).

Initially, the genome was probably unsegmented, that is, separate genes were on separate pieces of RNA (all genes unlinked), as is the case today in the simplest ssRNA viruses such as the influenza virus. Unsegmented genomes would facilitate recovery from genetic damage through "hypercyclic cooperation," a process in which an undamaged copy of a gene duplicates to make up for the damaged copy (Bernstein et al. 1984); no physical molecular recombination is required. Segregation of genes into daughter cells is sloppy in modern unsegmented viruses such as the influenza virus, and many defective phage particles are produced. ssRNA evolved into double-stranded RNA because a double-strand duplex is more stable. Reovirus is an example of a segmented double-stranded RNA virus that undergoes a kind of sexual repair termed multiplicity reactivation (McClain and Spendlove 1966).

As life continued to evolve, the number of genes increased and accurate segregation became critical. The segregation problem could be solved by linking genes together into a smaller number of linkage groups or chromosomes. By so doing there would be fewer daughter cells missing any necessary gene. This would have the further advantage of reducing the conflict among genes (Maynard Smith and Szathmáry 1993). A modern example of an unsegmented RNA virus is bacteriophage ø6, which has three duplex RNA segments containing several genes each. However, with linked genes a damage anywhere in the segment (in any particular gene) would invalidate the entire segment (all genes on the chromosome) unless a way could be found of repairing one or a few specific nucleotide sites. Molecular recombination with its accurate breakage and reunion of duplexes was the answer. DNA, with its standard base pairs, became the nearly ubiquitous genetic material, probably because of its greater stability and greater ease of molecular recombination (Cox 1991).

Bigger things eat smaller things and so there must have been a tendency for primitive living things to get larger. A single cell can get just so big, because the surface to volume ratio rapidly declines as cells get big. Surface area is needed to transport food and get rid of wastes. These are most likely the main reasons that multicellularity first arose. Although it is simple to imagine the initial advantage of multicellularity, there are difficult problems in understanding the transition from unicellular to multicellular life. Most critical is explaining how single cells relinquish their hold on fitness in favor of the new emerging unit, the multicellular organism. This is the topic of chapters 5 and 6.

## HERITABLE CAPACITIES OF SINGLE CELLS

The heritable capacities of the molecular replicator studied in chapter 2 fall into three basic categories, birth, death, and the utilization of resources (table 2-1). As life continued to evolve and diversify into higher-level units such as single cells, what classification of heritable capacities is appropriate?

To improve the phenotypic features underlying the capacities of birth, survival, and resource utilization, a certain degree of decomposability of traits is necessary. The advantage of cooperative interactions among traits must be balanced with a degree of independence, or else every trait ends up affecting every function, leaving little opportunity to improve one function without detracting from another. At some point traits must trade off against one another, however, and the study of life histories is replete with examples of traits that increase one component of fitness, say, fertility, while detracting from another, say, survival (consider the peacock's tail or other sexually selected traits). Still, there must be a balance between the interaction and decomposition of traits and their functions. What is unclear is how this decomposability of traits is produced during evolution.

Consider simple extant cell-like life forms such as viruses and bacteria. My colleagues and I have argued that for two of the best studied unicellular organisms in which gene functions are understood in some detail—the bacteriophage T4 and the bacterium *Escherichia coli*—the general heritable capacities $b_i, d_i,$ and $R_i$ continue to provide a natural categorization of the adaptive design for most of the specific molecular and supramolecular structures and processes encoded by the nucleotide sequence (Bernstein et al. 1983). Since we first analyzed the heritable capacities of these organisms, the genomes of T4 and *E. coli* have been completely sequenced and a more precise breakdown of gene function is possible. The story remains basically the same as it was in 1983 (H. Bernstein, personal communication).

For example, phage T4 is a virus that infects the bacterium *E. coli* and then the cell undergoes lysis producing several hundred phage progeny in each cycle. The virus genome has been completely sequenced and has about 308 genes, of which over half have been characterized with respect to function (Blattner et al. 1997). Of these genes, approximately 8% promote replication, 33% reduce mortality, 51% function in obtaining and processing resources, and 7% have uncertain roles.

For more complex organisms matters rapidly become more intricate. The complete genome sequence of the bacterium *E. coli* is now known (Kutter et al. 1994). It has about 4288 genes, of which about 2630 have been characterized with respect to function. These gene functions generally fall into the three categories of birth, survival, and resource utilization, although it is natural to subdivide each of these categories into subcategories. For example, with regard to the heritable capacities underlying the birth process, the genes are involved in either DNA replication or cell division. While with phage T4 the genes usually fell unambiguously into one of the three categories, in the case of the more complex bacterium some structures contribute to protection, that is, avoiding death, and obtaining resources. The vast majority of genes in *E. coli* are involved in resource utilization as indicated by the large number of enzymes involved in intermediary metabolism.

## Reconsidering Adaptedness and Fitness

Fitness emerged from the chaos when molecules became capable of self-replication. Template-mediated self-replication considered in chapter 2 is a Malthusian growth process (exponential) in which birth depends upon density in a linear fashion. Non-Malthusian growth (subexponential) for the template-mediated molecular replicator is also possible. In protein-catalyzed replication, as with sexual reproduction, birth depends upon density in a nonlinear quadratic fashion (superexponential growth). These different forms of replication affect the dynamic of selection in profound ways, giving rise to different characterizations of natural selection, as capsulized in the phrases "survival of the fittest," "survival of anybody," and "survival of the first." In survival of the fittest, the heritable capacities of the individual units of selection are the ultimate determinant of their evolutionary success. It is possible to combine the heritable capacities of the genotype or phenotype into an overall measure of fitness or adaptedness (recall equations 2-5 and 3-3). In survival of anybody, there is a cost to commonness and new types can invade when rare. This can be viewed as a diversity-producing effect, but also it may have a destabilizing effect on communities. In survival of the first, there is a cost of rarity in that heritable capacities have no role to play if a type must increase from low density, as must be the case for many mutants. Commonness alone can stabilize a community, regardless of the design attributes of the phenotype. In many ecological situations of interest, especially in structured populations, spatial matters become of overriding importance in determining whether a type may invade. There is simply not a property, or combination of properties, of the unit of selection that alone determines its evolutionary success. There is no adaptedness, in the sense of an overall property of an individual that alone predicts its evolutionary success.

## EARLY TRANSITIONS IN EVOLUTION

The transitions facing the early replicators were to form communities of cooperating replicators termed hypercycles, which eventually became encapsulated in a protocell. The evolutionary passage from genes to gene networks to cells has the components of later evolutionary transitions. These components include cooperation among lower-level units to produce a new higher-level unit, the resulting problems of frequency-dependent selection as defection and conflict arise, the evolution of individuality as a means to cope with conflict, and the reinvention of sex at the new higher level as a means to protect the well-being of the genes. New evolutionary units must find ways of promoting cooperative interactions while regulating the inherent tendency of lower-level units to compete. Ultimately, the new unit may be defined by the evolution of an encasing structure such as a cell membrane; however, before this occurs density and frequency dependence play a dominant role in the evolutionary dynamics. In an important sense individuality is a way of avoiding the limitations inherent in frequency-dependent selection. Because the advantages of interactions are achieved only at significant densities, cooperation and hypercycles have difficulty increasing when rare in populations of molecules replicating via template-mediated complementarity. These problems are compounded by any costs inherent in primitive translation systems. Two pathways are investigated here for increase of the hypercycle. The first is based on Eigen's quasispecies distribution. Here the hypercycle is assumed to hitchhike to significant density by virtue of its production as a mutant of a dominant master sequence replicating without enzymes via template means. The second way for the hypercycle to increase is by being localized in space because of either local diffusion or self-structuring of the local environment.

In the first chapter, I introduced some of the basic terminology of the theory of the evolution of interactions. Now is the time to pause in our study of evolutionary transitions and develop these tools in more detail so as to study the continued increase in complexity during the transition from single cells to multicellular organisms in later chapters.

# Evolution of Interactions

I HAVE already mentioned that cooperative interactions between units of selection are the basis of higher-level, more inclusive units. In this chapter, I develop a theory for the evolution of interactions using the methods of population genetics and game theory. This framework will be used in chapters 5 and 6 to consider the issue of individuality of evolutionary units and the passage of single cells into groups of cooperating cells (multicellular organisms).

## GENE FREQUENCY CHANGE

Consider a single gene locus with two alleles $A$ and $a$ in a diploid organism with absolute individual fitnesses defined as the expected number of gametes produced by each genotype (table 4-1). The fitnesses may be frequency and/or density dependent, but I am intentionally not being explicit about the ecological causes of the fitness differences for the time being. Throughout the book, I study the evolution of interactions between different types of individuals: molecules, cells, and organisms. These interactions are represented in terms of explicit models in which the underlying causes of the individual fitness differences are made explicit. In table 4-1, I assume random mating with Hardy–Weinberg frequencies at the start of the generation. The subscript used in table 4-1 is useful when summing terms and refers to the number of $a$ alleles in the genotype.

By counting the number of $a$ alleles at the start of the next generation the following gene frequency equation can be obtained:

$$\Delta q = \frac{pq}{2\overline{W}} \sum_i W_i \frac{df_i}{dq},$$   (4-1)

with the average fitness of the population (equivalent to the average fitness of individuals within the population) being given by

$$\overline{W} = \sum_i f_i W_i .$$   (4-2)

49

| Genotype | $AA$ | $Aa$ | $aa$ |
|---|---|---|---|
| Fitness | $W_0$ | $W_1$ | $W_2$ |
| Frequency | $f_0 = p^2$ | $f_1 = 2pq$ | $f_2 = q^2$ |
| Derivative | $\dfrac{df_0}{dq} = -2p$ | $\dfrac{df_1}{dq} = 2 - 4q$ | $\dfrac{df_2}{dq} = 2q$ |

*Table 4-1*

Single-locus selection model.

## POPULATION GROWTH

I use absolute individual fitnesses ($W_i$ defined as the expected *number* of gametes produced by individuals of genotype $i$) throughout this book, because I am interested in questions involving multiple levels of selection. Absolute fitnesses are required to partition fitness effects among the different levels of selection. Consider just a single group or population. By using absolute fitnesses we may study the growth rate of the population in addition to gene frequency change. The dynamics of population size, $N$, associated with gene frequency change (equation 4-1) can be shown to be

$$\Delta N = (\overline{W} - 1)N \tag{4-3}$$

(see, for example, Roughgarden 1979, chap. 3). Here, $\overline{W}$ is the average fitness of the population and is understood to depend on the gene and genotype frequencies in the population (equation 4-2). In unstructured populations with selection at the level of individual organisms, equations 4-1, 4-2, and 4-3 describe evolution. Because of the role of $\overline{W}$ in population growth, average individual fitness is often viewed as a measure of well-being or state of adaptation of the population. For this reason, we are very interested in the effects of evolution on the average fitness of the population.

So far, I have not assumed anything about the genotypic fitnesses except that they can be determined from the variables in the model if need be—either genotype frequencies in the case of frequency-dependent selection or population size in the case of density-dependent selection. The fitnesses could be constant or time varying; the basic selection equations above still apply. I have also assumed that the genotype frequencies may be obtained by the gene frequencies, through specification of the mating system; for the time being I have assumed random mating.

## FREQUENCY-DEPENDENT SELECTION

Different genotypes may behave in different ways. The effect of interactions among the genotypes may then depend upon the frequency of different types in the population. We say that fitness is then frequency dependent. When gene frequencies change, so must the frequencies of different genotypes, and this change will likely have some effect on fitness and the rate of growth of the population. Consider the rate of change of average fitness with gene frequency in the general case given in equation 4-4, where genotypic fitness may depend on gene frequencies:

$$\frac{d\overline{W}}{dq} = \sum_i W_i \frac{df_i}{dq} + \sum_i f_i \frac{dW_i}{dq}$$

$$= \sum_i W_i \frac{df_i}{dq} + \frac{\overline{dW_i}}{dq}.$$

(4-4)

The second term may be interpreted as the average of the current sensitivities of individual fitness to gene frequency. Equation 4-4 may be used to rewrite the basic gene frequency equation 4-1 as

$$\Delta q = \frac{pq}{2\overline{W}} \left( \frac{d\overline{W}}{dq} - \frac{\overline{dW_i}}{dq} \right).$$

(4-5)

Wright first expressed gene frequency change in the form of equation 4-5 (Wright 1931). The terms in front of the bracket are always non-negative; so an internal equilibrium requires the bracketed terms to sum to zero. Consequently, at equilibrium the effect of a change in gene frequency on average population fitness exactly balances the average of its effects on the genotypic fitnesses. By partitioning gene frequency change in this way, we see how the total change under frequency-dependent selection depends upon the effects at the two levels, the population and the individual, even though there is no population structure in the model yet. The roles of these different components of gene frequency change in selection in group-structured populations will be further studied below.

## CONSTANT SELECTION

If selection is constant then the second term in the bracket in equation 4-5 is zero and we have

$$\Delta q = \frac{pq}{2\overline{W}} \frac{d\overline{W}}{dq}.$$ (4-6)

Under constant selection average fitness controls both gene frequency change (equation 4-6) and population growth (equation 4-3). Further analysis of equation 4-6 shows that selection maximizes the average fitness of the population, and, consequently, the well-being of the population as measured by its growth rate.

Because gene frequency change can be written as a function of the slope of the average fitness function in equation 4-6, evolution can be seen as moving about on an adaptive landscape or topography: climbing out of valleys low in population fitness to reside on the peaks. The adaptive topography is a powerful metaphor for understanding evolution.

## Adaptive Topography

The adaptive topography concept can be generalized to cases of frequency-dependent selection by returning to the basic gene frequency equation 4-1 and defining $F(q)$ as Wright's "fitness function":
The indefinite integral in equation 4-7 may not always exist, but it does exist in many cases of interest concerning the evolution of interactions, for example, in kin selection, where it equals the average inclusive fitness effect

$$F(q) = \int dq \sum_i W_i \frac{df_i}{dq}.$$ (4-7)

Using equation 4-7 in equation 4-1 we see that the general form of the adaptive topography (allowing for frequency-dependent selection) is

$$\Delta q = \frac{pq}{2\overline{W}} \frac{dF}{dq}.$$ (4-8)

(Michod and Abugov 1980). However, the satisfying connection between the gene frequency dynamics and population growth existing under constant selection no longer holds. When interactions are important, there is no simple connection between the evolution among genotypes within populations (or groups) and the well-being of the population (or group) as measured by the average fitness of the individuals within the population.

This characteristic of frequency-dependent selection poses a problem for the evolution of higher-level units which must originate as groups of lower-level units (gene networks, multicellular organisms), because selection at lower levels need not benefit the higher level. Cooperation benefits the higher level but frequency-dependent effects make cooperation vulnerable to defection. So it is no surprise that the culminating stage in the transition to a new unit of selection involves institutionalizing cooperation by removing its frequency dependence, as, for example, when the cell structure evolved to manage the costs and benefits of protein production (chapter 3). A similar result will be shown to hold for the evolutionary passage from cells to multicellular organisms (chapter 6).

Using the fitness function defined in equation 4-7, Wright derived the version of Fisher's fundamental theorem (see the section "Fisher's Fundamental Theorem" below) for frequency-dependent selection given in equation 4-9:

$$\Delta F \cong \text{Var}[\overline{W}_i], \tag{4-9}$$

where $F$ is defined by equation 4-7 and $\overline{W}_i$ is the marginal fitness of allele $i = A,a$ for the simple one-locus two-allele case under consideration here (see Wright 1977, pp. 422–423). The problem with frequency-dependent selection is (again) that the fitness function defined by equation 4-7 bears no obvious relation to the well-being of the group or the average fitness of individuals comprising the population (compare equation 4-9 with equation 4-15 below). The main advantage of equation 4-9 is that the fitness function is dynamically sufficient (through equation 4-8), while Fisher's representation in equation 4-15 below is not dynamically sufficient (Frank and Slatkin 1992), because the gradient of average individual fitness does not describe natural selection when interactions are important.

## FREQUENCY DEPENDENCE DECOUPLES FITNESS IN A SELECTION HIERARCHY

Wright proposed a grand scheme termed the shifting-balance process for integrating the effects of selection among individuals within populations (phase II) with the effects of selection between subpopulations in a larger ensemble of subpopulations (phase III; see Wright 1977, chap. 13). In the final phase (phase III) of Wright's shifting-balance process of evolution, local populations with especially advantageous combinations of traits become hot spots for migration and force their local constellation of genes into the global population. This asymmetric diffusion (as Wright called it) from especially well-adapted locales may lead to a transformation of the entire species, if the traits are adaptive over the entire range of the species. Wright viewed a partially subdivided population as the most adventitious for continued evolution of the species. In a partially subdivided population, local adaptation is possible and so is mass transformation of the species. Let us examine this idea a little further, to see if it may provide us with the framework we need to understand evolution in group-structured populations and the passage between lower and higher evolutionary units.

The average fitness of the local population describes its output into the global population through equation 4-3. When fitness is constant, the average fitness of the local population also controls within-population change among individuals through equation 4-6. Consequently, for constant selection, there is a pleasing harmony between selection at two levels in a selection hierarchy—here, the organism and population (or group) levels. Traits that increase the fitness of individuals also increase the average well-being of the group, and this may lead to a transformation of the entire species or global population. When selection is constant, there is no conflict among the two levels in the hierarchy, and the shifting-balance process Wright envisioned can operate, with the third phase of group selection building upon the gene frequency change occurring during the second phase of within-group selection.

Frequency-dependent effects on fitness changes all this. The fitness function $F(q)$ given in equation 4-7 controls local evolution in subpopulations through equation 4-8. However, local population growth is still controlled by average individual fitness $\overline{W}$ through equation 4-3. There need be no necessary relation between the two functions ($\overline{W}$ and $F(q)$). In frequency-dependent models, within-population change among organisms can lead to demise of the subpopulation and even local extinction (for some simple examples, see Wright 1969). This tension between the well-being of the group (or subpopulation) and selection among its components is the basic problem that must be solved during evolutionary transitions in the units of selection. The problem is nonexistent when selection is constant, and in this case the two phases of the shifting-balance process operate in concert, as the phase of asymmetric diffusion builds upon the phase of within-subpopulation change. However, when fitness depends upon interactions, the well-being of the group is threatened by the evolution within. The harmony between the two levels existing under constant selection is lost, and the phase of asymmetric diffusion can no longer build upon the phase of within-subpopulation change in a concerted manner.

Frequency dependence and the decoupling of fitness levels in a selection hierarchy have not been considered in recent analyses of Wright's shifting-balance process (Barton 1992; Coyne et al. 1997; Crow et al. 1990; Gavrilets 1996; Kondrashov 1992). Frequency dependence would seem to present a fundamental problem to the third phase of the shifting-balance process, at least if the third phase is expected to build upon the changes occurring during the second phase.[1] However, one can interpret Wright's shifting-balance theory in a more general light, as a set of interactive processes occurring at different levels, without any expectation that selection will proceed in the same direction at the different levels. This view is recommended by the fact that Wright himself often studied frequency-dependent selection; he understood well the problems posed by it and the fact that frequency-dependent selection often does not increase mean (group) fitness. Indeed, one can find in Wright's structured population model a mechanism, namely, group selection, that can inhibit the tendency of within-population selection to favor defection and decrease mean (group) fitness.

In any event, other analytical tools are needed to understand evolutionary transitions in multilevel settings when the fitnesses at different levels depend upon interactions. Three approaches are introduced now to be used in later chapters. The first is the covariance approach to selection introduced by R. A. Fisher and later developed by G. Price and A. Robertson. The second approach involves the methods of kin selection in the study of evolution in genetically structured populations. The third approach involves the methods of game theory and its study in structured populations.

## Selection as Covariance

I introduced the covariance approach to selection in chapter 1 without any special mention of genetic selection, which I consider now in the context of Price's approach. By viewing fitness as a quantitative trait, much like body weight or milk yield, selection in multilevel settings may be studied using statistical methods. The representation of selection as a covariance (equation 1-7, or equation 4-10 below, or equivalently the regression equation 4-11) of fitness and individual gene dosage (either individual gene frequency or gene number) was independently derived by several authors (Li 1976, 1967a,b; Price 1995, 1972a, 1970; Robertson 1968, 1966). However, as discussed in more detail below, Fisher was the first to make use of covariance approaches to selection in his definition of the "average effect" of an allele in the derivation of his fundamental theorem of natural selection (Fisher 1941, 1930, 1958).

$$\Delta q = \frac{\text{Cov}[W_i, q_i]}{\overline{W}}, \qquad (4\text{-}10)$$

$$\Delta q = \frac{\text{Reg}[W_i, q_i]\,\text{Var}[q_i]}{\overline{W}}. \qquad (4\text{-}11)$$

The individual gene frequency in equations 4-12 and 4-13 below depends on gene dosage and ploidy level and is given for any particular type $i$ as $q_i = g_i/n_z$, where $g_i$ is the gene dosage in individual $i$ and $n_z$ is the zygotic ploidy of the species concerned. An equation similar to equation 4-10 can be developed for a trait determined by quantitative genetic variation. In this case the selection response is expressed in terms of the heritabilities and correlation of fitness with breeding values for the trait (see Falconer 1989, pp. 318–322). George Price's derivation of equation 4-10 is straightforward (Price 1970). We simply consider the transformation of the offspring of one generation into the offspring of the next. Enumerate all the individuals of the first generation with the index $i$ where $i = 1, 2, \ldots, N$ and $N$ is the total number of individuals. Two variables figure in the analysis of the covariance; they are $q_i$, the individual gene frequency within individual $i$, and $W_i$, the fitness of individual $i$ (the expected number of gametes produced). The gene frequency among the first generation, $q$, is given by

$$q = \frac{\sum_i q_i}{N}. \qquad (4\text{-}12)$$

The gene frequency at the start of the next generation is given in equation 4-13, where $\overline{W}$ is the average fitness of the individuals in the first generation:

$$q' = \frac{\sum_i W_i q_i}{N\overline{W}}. \qquad (4\text{-}13)$$

By subtracting the new gene frequency from the old (and using the definition of covariance) we get equation 4-10. One may verify that for a single locus with two alleles equation 4-10 equals equation 4-1 (see below).

The covariance approach is especially well suited to studying selection at multiple levels. Equation 4-13 assumes that gene frequencies within individuals do not change. However, during development, mutation and selection may act to change the frequencies of cells within organisms and such changes may be included, as discussed below (see equation 4-16, figure 4-1, and table 4-4 below).

I now relate the covariance approach to the selection at a single diploid locus considered at the beginning of this chapter. The index $i$ refers to the different kinds of genotypes (say, *AA, Aa, aa* for a single diploid locus with two alleles), $q_i$ is a property of the genotype such as the dosage of a particular gene (say, the $a$ gene, so that $q_{AA} = 0$, $q_{Aa} = \frac{1}{2}$, $q_{aa} = 1$, and $f_i$ is the frequency of genotype $i$. In discrete, nonoverlapping-generation models (the standard model of population genetics; see, for example, table 4-1 above) the unselected set is the population of offspring in one generation and the selected set is the population at the same stage in the next generation. The average population properties (given in equations 1-3 and 1-4 or equations 4-12 and 4-13) are simply the gene (or genotype) frequencies in the two successive generations.

Because of sexual recombination, parents may produce offspring with different genotypes from their own. As a result, homozygous *AA* offspring may come from gametes produced by *AA* or *Aa* parents. Consequently, the transformation through time embodied in equation 1-5 no longer involves only the properties of $i$ types. With sex, the frequency of a particular genotype in the next generation depends on the properties of other genotypes along with the mating and genetic systems. This is the property change mentioned in the first chapter when I discussed the differences between natural selection and everyday selection (see table 1-1).

The transformation envisioned in equation 1-5 is becoming complex, so let us partition the life cycle into manageable steps. After survival and production of gametes (before mating and fusion of gametes) there is a natural break in the life cycle (at least for the model organism). So let us use the concept of individual fitness, $W_i$, defined as the expected number of gametes produced by type $i$. With this concept of fitness, the change in gene frequency is still described by equation 1-7 (Price 1970), with the slight modification that the right-hand side is divided by $\overline{W}$. This is necessary because the average fitness is no longer unity for the case of genotypic fitness, as it is if fitness is defined by equation 1-5. We then have $\text{Cov}[q_i, W_i] = \sum_i f_i q_i W_i - \overline{W}q = f_{aa}W_{aa} + \frac{1}{2}f_{Aa}W_{Aa} - \overline{W}q$. Plugging this last expression into equation 1-7 and dividing by $\overline{W}$ gives the standard equation for gene frequency change: $\Delta q \overline{W} = f_{aa}W_{aa} + \frac{1}{2}f_{Aa}W_{Aa} - q\overline{W}$.

## FISHER'S FUNDAMENTAL THEOREM

The regression of individual fitness on individual gene frequency studied in the last section (equation 4-11) may be used to predict the effect of gene frequency change on the average fitness of the population (Fisher 1941). Consider a population of organisms composed of different genotypes. For each organism, we measure its individual fitness (expected reproductive success) and its genotype at a single diploid locus of interest (say with two alleles $A$ and $a$) as represented by its individual frequency of allele $a$, $q_i =$ 0, 1/2, 1, for genotypes $AA$, $Aa$, and $aa$, respectively. Imagine plotting all organisms in the population on a graph, the abscissa being individual frequency, $q_i$, and the ordinate being individual fitness. For any real population, there will be a distribution of points plotted for each $q_i$, the density of points depending upon the genotype frequencies in the population $P$, $2Q$, and $R$ for $i = AA$, $Aa$, and $aa$, respectively. The variance of fitness for each genotype class will depend on environmental and random factors as well as the canalization of fitness. In theory, we may consider the fitnesses of each genotype to be concentrated at a single value, $W_i$, with weights $P$, $2Q$, and $R$ for $i = AA$, $Aa$, and $aa$, respectively. In practice, genotypic fitness $W_i$ will be an average value, but we ignore this complication, following Fisher (1941).

As the population evolves, the frequencies of the genotypes change, as do the individual fitnesses $W_i$ (if there is frequency dependence), while the individual gene frequencies, $q_i$, remain fixed at their discrete values. For the purpose of understanding the effects of gene frequency change on fitness, Fisher (1930, 1941, 1958) defined $\alpha$, the average effect of a gene substitution, as the slope of the linear regression fitted to these points by the technique of least squares, using the population genotype frequencies as weights (table 4-2). Fisher's approach embodies the covariance methods introduced above (for example, equation 4-11), since by definition $\alpha = \text{Reg}[W_i, q_i]$.[2] At any set of genotype frequencies, the fitnesses of the three genotypes can be predicted by

| | Individual Fitness | | Population | Individual |
| Genotype | Actual | Predicted By Additive Model | Frequency of Genotype | Frequency of $a$ Allele |
|---|---|---|---|---|
| $AA$ | $W_{AA}$ | $W_{AA}^+ = \mu - \alpha/2$ | $P$ | 0 |
| $Aa$ | $W_{Aa}$ | $W_{Aa}^+ = \mu$ | $2Q$ | $\frac{1}{2}$ |
| $aa$ | $W_{aa}$ | $W_{aa}^+ = \mu + \alpha/2$ | $R$ | 1 |

Table 4-2

Fisher's average effect of gene substitution. *Source:* Fisher 1941.

the additive model given in table 4-2, where $\alpha/2$ is the additive effect on individual fitness of changing individual gene frequency by amount 1/2.

Fisher first stated his fundamental theorem of natural selection (1930, 1958) as "The rate of increase in fitness of any organism at any time is equal to its genetic variance in fitness at that time." Later, Fisher (1941) restated his theorem as "The rate of increase of the average fitness of the population is equal to the genetic variance of fitness of that population." There are many derivations of Fisher's theory. Using the regression form of the gene frequency, equation 4-11, we may derive Fisher's fundamental theorem of natural selection as follows. From the very definition of regression of fitness on gene frequency (thinking in terms of the plot of individual fitness on gene frequency mentioned above), we may predict on the basis of the additive model of fitness that a small change in population gene frequency, $\Delta q$, has the following effect on the average fitness of individuals:

$$\Delta \overline{W} = \alpha \Delta q \qquad (4\text{-}14)$$

(see Fisher 1941, p. 57). Upon substituting $\Delta q$ from equation 4-11 into the right-hand side of equation 4-14 (and assuming weak selection so that the denominator of the right-hand side of equation 4-11 $\overline{W} \approx 1$), we get Fisher's theorem in the form

$$\Delta \overline{W} = \alpha^2 \text{Var}[q_i] = \text{Var}[W_i^+]. \qquad (4\text{-}15)$$

Using the construction given in table 4-2, it can be seen that the genetic variance in fitness assuming the additive model, $\text{Var}[W_i^+]$, is simply the middle expression in equation 4-15, $\alpha^2 \text{Var}[q_i]$.[3]

There has been much discussion concerning what Fisher's fundamental theorem is supposed to mean (Edwards 1990; Frank and Slatkin 1992; Li 1967a; Price 1972b). The most complete analysis makes clear that Fisher's theorem applies to the change in average fitness caused by selection, assuming that "environment" is held constant (Price 1972b). By the "environment" Fisher meant not just the biotic environment (such as competitors, predators, or cooperators) and physical environment (such as temperature) but also the genetic environment (such as nonadditive gene effects, dominance, and changes in the regression of fitness on individual properties). By "natural selection" Fisher meant selection of additive gene effects in a constant environment. In Price's words,

What Fisher's theorem tells us is that natural selection (in his restricted meaning involving only additive effects) at all times acts to increase the fitness of a species to live under the conditions that existed an instant earlier.

But since the standard of "fitness" changes from instant to instant, this constant improving tendency of natural selection does not necessarily get anywhere in terms of increasing "fitness" as measured by any fixed standard, and in fact $M$ [average fitness] is about as likely to decrease under natural selection as to increase.

Why was Fisher so interested in selection of additive gene effects in a constant environment? There are two reasons. The first can be found in Fisher's work and has been discussed by subsequent workers (Frank and Slatkin 1992; Price 1972b). Fisher wanted to say something general about natural selection and felt that his theorem was the most that could be said. I think he must have also felt that it is hard to say anything general about the deterioration of the environment, except that most of the time it will undo any progress made by selection. Furthermore, Fisher often stated that the total change in average fitness of a population (the change in average fitness caused by natural selection plus the change in average fitness caused by the environment) had to hover around zero most of the time for populations in equilibrium.

The second reason I think Fisher partitioned evolution in this way has not been explicitly acknowledged in the literature, as far as I know, perhaps because it is so simple. Fisher's theorem expresses the commonsense notion of what we expect from everyday selection discussed in chapter 1 ("Selection as Fitness Covariance"). If I go to the grocery store and select items according to low cost, I expect the average cost of items in my grocery basket to be lower than if I had chosen the items at random. This is basically what Fisher's theorem expresses, except the selected criterion is fitness instead of cost. After selection on the basis of individual fitness, we expect the average fitness of individuals to increase, as Fisher's theorem states. As discussed in the first chapter (table 1-1), the commonplace notion of selection ignores many of the special features of natural selection. Because Fisher's theorem basically restates in biological terms the commonsense notion of selection, I agree with Price that it is "less important than he [Fisher] thought it to be" in understanding *natural* selection.

Recall Price's (1995) schema for natural selection discussed in "Selection as Fitness Covariance" in chapter 1, which includes everyday "selection" along with "property change" (resulting from mating, sexual recombination, and other aspects of genetics). Under natural selection the chosen parents (say they are heterozygotes, *Aa*), produce offspring with different properties (like those of the two homozygotes, *AA* and *aa*), because heterozygotes cannot breed true when there is sex. It is as if, after selecting the inexpensive brands at the grocery store, I get home to find the expensive kind of toothpaste in my bag! It doesn't sound like everyday selection, and it isn't, but natural selection often works this way when the mechanics of mating and heredity are taken into account.

Frank and Slatkin (1992) give an example of the evolution of clutch size based on Cooke et al. (1990), in which the total change in clutch size can be partitioned into selection in a constant environment and the response of the environment to changes in clutch size (resulting from increased competition for space). Frank and Slatkin's analysis is satisfying precisely because it gives a full description of the process of natural selection on clutch size, including the systematic ecological processes involved. I think that Frank and Slatkin are being too generous in attributing to Fisher their own insights, along with those of Cooke et al. (1990), about the explicit role of ecological dynamics in natural selection.

Much progress has been made since Fisher's day in deconstructing the environmental component of fitness. Indeed, I see the work on fitness deconstruction mentioned in chapter 1 (especially the many within- and between-species ecological models), and the models of frequency-dependent interactions discussed in the present chapter, as directed toward the general goal of having a dynamically sufficient theory of all the aspects of natural selection relevant to the problem at hand, including not only genetic change but also the systematic (nonrandom) components of environmental change. In the quest for dynamical sufficiency, certain well-engrained attitudes about selection based on everyday experience must be abandoned.[4]

Consider Wright's version of Fisher's theorem for frequency-dependent selection given in equation 4-9 above. From what I have said concerning Fisher's goal in separating out selection in a constant environment from changes in the environment, Fisher would probably not have agreed with Wright's associating equation 4-9 with his (Fisher's) theorem. The left-hand side of Wright's equation 4-9 includes the environment in "fitness." However, "fitness" is no longer the average individual fitness but rather Wright's "fitness function" given in equation 4-7. Wright's equation 4-9 is a dynamically sufficient description of all the changes occurring in the model, including changes in the environment (in this case, the environment is the frequency of genotypes in the population). Wright accomplishes this by ignoring the very distinction Fisher wanted to make, and by redefining fitness so that it no longer measures the well-being of the individuals or the group. Wright's fitness function may have an intuitive interpretation, as in the case of the average inclusive fitness effect when the environment includes genetic relatives (discussed below equation 4-19), but Wright's fitness function still bears no necessary relationship to population growth and the average well-being of individuals.[5]

Frequency-dependent interactions between units are the basis of new higher-level units. Yet one of the most basic consequences of frequency-dependent natural selection is that there is no necessary progress made in terms of the well-being of the individuals involved or in terms of the group. How can frequency-dependent interactions be the basis of higher-level units but not lead to the increase in fitness of these new units? This paradox of frequency dependence is the basic problem that must be solved during the transition to a new unit of selection. To understand the paradox more fully, and its eventual resolution, I now extend the covariance methods of selection to hierarchically structured populations. These methods will be used in later chapters to study the emergency of individuality.

## EVOLUTION IN HIERARCHICALLY STRUCTURED POPULATIONS

In the development of the covariance form of selection given in equation 4-10, the individual units of selection (we had in mind individual organisms) were well defined. An advantage to Price's model is that once we pick a focal level it is straightforward to include higher or lower levels (Price 1970, 1972a; Hamilton 1975). For example, we can include forces like mutation, meiotic drive, and gene conversion by including terms for the change in gene frequency within organisms. In this case, the individual gene frequency, $q_i$, in equation 4-13 becomes a function of processes occurring within the life span and development of the organism. Or we could look to changes involving groups of organisms in structured populations if we were interested in the evolution of social behaviors. In either case the methodology is the same. For each unit, we need to consider "weights," regarded as absolute fitnesses, that map the frequencies of the units (molecules, cells, organisms, populations) from one appropriate time unit to the next. The time units are chosen so that the defined processes may be applied in a recursive manner.

Consider selection among organisms in groups and between groups within a total population (table 4-3). Proceeding as before, I enumerate organisms within groups with the subscript $i$ and the groups themselves by the subscript $s$. Let the number of organisms in group $s$ be $N_s$. Since groups reproduce by virtue of the reproduction of organisms, the fitness of the group, $\overline{W}_s$, is taken as the average of the fitnesses of its member organisms, so that

| | Organism ($i$) | Group ($s$) | Total Population |
|---|---|---|---|
| Number of Genes | 2 | $2N_s$ | $N = 2\sum\limits_s N_s$ |
| Frequency | $q_{i,s} = 0, \frac{1}{2}, 1$ | $q_s = \dfrac{1}{N_s} \sum\limits_{i=1}^{i=N_s} q_{i,s}$ | $q = \dfrac{1}{N} \sum\limits_s N_s q_s$ |
| Fitness | $W_{i,s}$ | $\overline{W}_s = \dfrac{1}{N_s} \sum\limits_{i=1}^{i=N_s} W_{i,s}$ | $\overline{W} = \dfrac{1}{N} \sum\limits_s N_s \overline{W}_s$ |
| New Frequency | $q_{i,s}'$ | $q_s' = \dfrac{1}{N_s \overline{W}_s} \sum\limits_{i=1}^{i=N_s} q_{i,s} W_{i,s}$ | $q' = \dfrac{1}{N'} \sum\limits_s N_s' q_s'$ |
| Change in Frequency | $\Delta q_{i,s} = 0$ | $\Delta q_s = q_s' - q_s$ | $\Delta q = q' - q$ |

*Table 4-3*

Selection in hierarchically structured populations. The variables $q_{i,s}$ and $W_{i,s}$ are the individual frequency and fitness, respectively, of individual $i$ in subpopulation $s$. Note that the frequency of each individual in subpopulation $s$ is $1/N_s$. Diploid individuals are assumed.

the number of offspring produced by group $s$ is $N_s' = \overline{W}_s N_s$. For similar reasons, the gene frequency of the offspring of a group, $q_s'$, is taken as the average gene frequency of the offspring of its members. Using the construction given in table 4-3, the gene frequency change in the total population, $\Delta q$, can be expressed as

$$\Delta q \overline{W} = \text{Cov}[\overline{W}_s, q_s] + E[\overline{W}_s \Delta q_s] \qquad (4\text{-}16)$$

(Price 1970, 1972a).

There are two components on the right-hand side of equation 4-16. The first is a covariance between fitness and gene frequency at the group level. The second term is an average of the changes in gene frequency occurring within groups. I use this equation in the next two chapters, when I consider the evolutionary passage from cells to multicellular organisms, but it may be worth pointing out some obvious virtues now. The equation partitions the total change in frequency into changes between and within groups. For a group to emerge as a new unit of selection, ways must be found to increase the covariance in fitness at the group level and to reduce the change within groups that results from selection among lower-level units. This will become especially useful in our investigations into the emergence from cells of groups of cells, or organisms, as new units of selection.

## EVOLUTION OF MULTICELLULAR ORGANISMS

Organisms may be thought of as groups of cooperating cells related by common descent. Selection among cells—below the level of the organism—could destroy this harmony and threaten the individual integrity of the organism. This competition could favor defecting cells that might pursue their own interests at the expense of the organism. For the organism to emerge as an individual, or unit of selection, ways must have been found for regulating the selfish tendencies of cells and promoting cooperative interactions among cells. These questions will be addressed in some detail in the next two chapters; for now I am only concerned with the basic framework involved and its relation to the general population genetic theory for the evolution of interactions.

Many organisms originate from groups of cells produced by budding (*Hydra*), by vegetative means (many plants), or by aggregation of single cells (slime molds). The hierarchical framework of Price (table 4-3) is applied to this situation in table 4-4. We begin with cells for propagules of type $i$. In a model currently under study, we assume haploid cells with two kinds of cell behavior, cooperate and defect (Michod and Roze 1998). There is a fixed initial propagule size $k$ for all types $i$, made up of $i = 1, 2, \ldots k$ cooperating cells and $k - i$ defecting cells. Using a mutation selection model of cell replication (similar to that studied in figure 5-1 and table 5-3 below), we calculate the distribution of cell types in the adult stage. Fragmentation of the adult occurs into propagule groups again of size $k$ (with the composition of each propagule $i$ calculated according to binomial sampling from the adult). Output of adult groups into the propagule pool increases with the number of cooperating cells in the group and this represents selection at the organism level (similar to that assumed in table 5-2 below). In the case of reproduction via single cells which aggregate into propagules (as in the case of *Dictyostelium discoideum*) a different reproduction and sampling scheme is used. In this way recurrence equations for the frequency distribution of propagule types $i$ may be generated.

Price's model as applied to organism-like cell groups (table 4-4) makes use of "weights" (the $k$ variables in table 4-4) that are assigned to classes (types) before and after selection. The classes are the various propagule types that start development. The weights are used in the model to quantify

| | Cell | Organism | Population |
|---|---|---|---|
| Number of Alleles | 2 | $2k_i, 2k_i'$ | $2\sum_i k_i, 2\sum_i k_i'$ |
| Frequency of $a$ Allele | $0,\frac{1}{2},1$ | $q_i, q_i'$ | $q = \dfrac{\sum_i k_i q_i}{\sum_i k_i}, q' = \dfrac{\sum_i k_i' q_i'}{\sum_i k_i'}$ |
| Fitness (Transformation) | $b$ | $\omega_i = \dfrac{k_i'}{k_i}$ | $\bar{\omega} = \dfrac{\sum_i k_i \omega_i}{\sum_i k_i}$ |
| Change in Frequency | | $\Delta q_i = q_i' - q_i$ | $\Delta q = q' - q$ |

*Table 4-4*

Group selection model of multicellular organisms. The subscript $i$ indexes kinds of organisms as determined by the type of propagule group. The table provides the basic framework for a hierarchical model of cell group that originate from propagule groups of size $k_i$ (by budding or other means such as aggregation). See text for explanation of how the variables $q_i$, $k_i$, $q_i'$, and $k_i'$ may be obtained. Diploid organisms are assumed. The parameter $b$ is the replication rate advantage of $aa$ cells relative to $AA$ cells (heterozygote cells replicate at some other rate depending on dominance).

the outcome of selection and other nonselective processes (such as mutation) that are involved in the transformation of the classes through time. In the present context the weights are the numbers of cells in the newly conceived propagule, $k_i$ (assumed equal for all $i$ in the above model), and in the gametes produced by the adult form, $k_i'$. A prime indicates a statistic or variable after the operation of selection and other processes that transform the frequencies and characteristics of the units under study. A feature of this model is that it easily handles different forms of mitotic reproduction such as budding and aggregation as just mentioned ($k_i > 1$) or zygote reproduction ($k_i = 1$). I now restrict my attention to zygote reproduction so that $k_i = 1$ for all $i$. In this case, there is no variable needed for propagule size and the prime notation is not necessary. It is then understood that $k_i$ is the number of cells in the adult.

An overview of a zygote model life cycle is given in figure 4-1. The subscript $j$ now indexes types of zygotes. After zygote formation (by either asexual or sexual means), cells proliferate during development to produce the adult form. This proliferation and development is indicated by the vertical

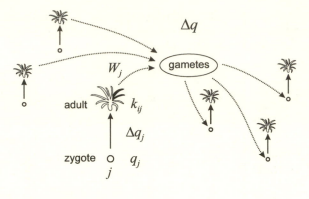

*Figure 4-1*

Model life cycle of organisms. Reproduction through a single-cell zygote stage is assumed. Adapted from Michod 1997a.

arrows in figure 4-1. Because of mutation and different rates of replication and death of the different cell types, the gene and genotype frequency of cells change within organisms, as represented by $\Delta q_j$ in figure 4-1. There will also be a change in gene frequency in the population due to differences in fitness between the adult forms. These two components of frequency change, within organisms and between organisms, give rise to the total change in gene frequency in the population, $\Delta q$. This framework and notation will be used in chapters 5 and 6 to study the evolution of individuality, including the function of the germ line and self-policing of cells by the organism.

## Kin Selection

Individual organisms are an example of kin-group structure; in this case, the groups are composed of cells clonally derived from a common zygote. Another common form of kin-group is family groups of organisms, because many offspring spend time in family units before dispersing. Wade has shown how covariance methods may be applied to selection in family-structured populations (Wade 1978, 1979, 1980, 1985), and in chapter 6 I use similar methods to understand the transition from cells to multicellular organisms. The adaptive topography approach yields similar results as the family-structured approach, with the advantage that it illuminates the role of inclusive fitness in selection in family- or kin-structured populations (Michod 1982; Michod and Abugov 1980; Uyenoyama and Feldman 1981).

Population structure is important because it constrains the distribution of interactions to be with genetically related types. We saw in our studies of the transition between hypercycles and cells in chapter 3 (especially equation 3-10) how population structure changes the experienced frequencies of types away from that expected if interactions occur at random in the total population. This is one way to overcome the limits posed by frequency-dependent selection. We may generalize this approach by defining genotypic fitness as a conditional expectation of interaction-specific fitnesses. Let $w_{ij}$ be the fitness of type $i$ when interacting with type $j$. Further, let $P(j/i)$ be the probability of interacting with type $j$ given that an individual is of type $i$ (identical to the conditional experienced densities used in our study of the evolutionary transition from genes to cooperative gene networks in chapter 3). The overall genotypic fitness for use in table 4-1 and equation 4-1 may be expressed as the conditional average of the association-specific fitnesses:

$$W_i = \sum_j w_{ij} P(j/i). \tag{4-17}$$

Of course, genotypic fitness calculated as a conditional average in the manner of equation 4-17 is dependent on gene and genotype frequencies.

What remains is to calculate the conditional distribution of interactions for various kinds of population structures. The methods for doing so in some common cases are explained elsewhere (Michod 1982; Michod and Hamilton 1980; Michod and Abugov 1980; Michod 1979). For simplicity consider an interaction with additive fitness effects:

$$w_{ij} = 1 + c_i + b_j. \tag{4-18}$$

In the case of full-sib family groups and weak selection (so that $F(q)$ in equation 4-7 is a function of $q$ alone), it can be seen (Michod and Abugov

$$F(q) = \sum_i f_i(c_i + \frac{1}{2}b_i), \tag{4-19}$$

1980) that Wright's fitness function defined in equation 4-7 equals the following average inclusive fitness effect:
where $f_i$ is defined in table 4-1.

Inclusive fitness, first defined by Hamilton (Hamilton 1963, 1964a,b), plays the role of the adaptive topography in equation 4-8. Evolution may be seen producing behaviors that maximize their inclusive fitness effect. But what is inclusive fitness?

Inclusive fitness is useful in understanding the outcome of natural selection in populations in which relatives are present. This means the concept is especially relevant to human evolution which, for the most part, has occurred in small groups of extended family members. The concept of inclusive fitness, and the attendant cost/benefit rule for spread of an altruistic gene, were developed by W. D. Hamilton (Hamilton 1963, 1964a,b) and have revolutionized the way biologists and some social scientists view behavior. In addition, inclusive fitness and Hamilton's rule (equation 4-22 below) provide a formal framework for many empirical studies on the evolution of behavior.

Relatives share genes. That is, because they have at least one common ancestor (parents, grandparents, great-grandparents, etc.), relatives possess copies of the same DNA sequence. These shared genes are termed "identical by descent," since they are identical by virtue of the fact that they are copies of a DNA sequence in a common ancestor and, therefore, are descended from that ancestor. In the simple but common case of a population structured into family units, different siblings carry genes that are identical by descent from their parents. In sexually reproducing populations, copies of an ancestor's genes exist in different individuals to varying degrees. Inclusive fitness takes this complex issue into account in predicting how interactions between individuals should evolve. For discussion, the individuals involved in an interaction are often called the "donor" and "recipient" of the behavior of interest.

Inclusive fitness thinking is based on the realization that an individual's offspring, say, the donor's offspring, are simply one kind of relative. Collateral relatives, say, siblings or first cousins, also carry genes that are identical with those in the donor. So these other kin can serve as vehicles for the perpetuation of the donor's genes, just as do the donor's offspring. Because of sexual reproduction, the donor's offspring carry approximately one-half of the donor's genes. Other relatives may carry an equally substantial portion of the donor's genes. For the same reason that the donor can pass on more of its genes by helping its offspring (for example, by providing parental care), so may the donor increase the frequency of its genes by helping these other relatives.

In population genetics, inclusive fitness begins with a baseline individual fitness, to which the different effects of the behavior are added. It is important to realize that the baseline fitness is the individual fitness that the donor would have without any social interactions. (This baseline fitness is usually assumed to be standardized to one.) The effects of the donor's behavior considered in inclusive fitness are the effects of the behavior on the donor's and recipient's baseline fitnesses, where the effects on the recipient's fitnesses are weighted by the degree of genetic relatedness of recipient and donor, $R$. (The relatedness of the donor to itself is, of course, one). Assuming the donor of interest is of genotype $i$ and using the additive model given in equation 4-18 ($c_i$ and $b_i$ are the additive effects of the donor's behavior on itself and the recipient), the inclusive fitness effect of $i$'s behavior is simply

$$c_i + Rb_i. \tag{4-20}$$

Inclusive fitness has three important functions in the study of natural selection in populations in which relatives are present. First, as we have seen, genes affecting behavior can be shown to increase, or decrease, according to whether their inclusive fitness effect is positive, or negative, respectively. In other words, for $i$'s behavior to spread,

$$c_i + Rb_i > 0. \tag{4-21}$$

An altruistic behavior is one in which the effect on the fitness of the donor is negative and the effect on the fitness of the recipient is positive ($c_i < 0$ and $b_i > 0$). Requiring the inclusive fitness to be positive (equation 4-21) for an altruistic behavior gives Hamilton's rule (dropping the $i$ subscript):

$$\frac{-c}{b} < R \tag{4-22}$$

for spread of the altruistic gene. Thus, the concept of inclusive fitness and Hamilton's cost/benefit rule are intimately related. We have already encountered a condition similar to Hamilton's rule in the study of the evolution of hypercycles in structured populations (equation 3-12).

The second important function of inclusive fitness is that, as we have seen, the average inclusive fitness effect in the population increases, and is eventually maximized, by natural selection. Finally, the inclusive fitness effect is a simple and straightforward quantity, depending as it does on just three parameters, $c$, $b$, and $R$. The most attractive aspect of inclusive fitness is its apparent simplicity. It summarizes the qualitative outcome of selection in a complicated frequency-dependent situation, by redefining fitness to include an important aspect of the environment, the relatedness of interactants.

## GAME THEORY

A widely used approach to studying natural selection, especially when there are interactions in fitness, is to liken life's contingencies to a game in which the interactions affect the fitness of the players involved. Game theory is another shortcut to describing the expected outcome of frequency-dependent selection. It is usually not dynamical in its approach (although haploid asexual dynamics can be assumed), instead game theory tries to capture the expected stable states of the system without considering whether these states are dynamically attainable.

The evolutionarily stable strategy, or ESS, is a state that cannot be invaded by rare mutants (Maynard Smith 1982; Maynard Smith and Price 1973). Parker and Maynard Smith (1990) suggest that the ESS should be considered as a kind of optimization approach to evolution, termed "competitive optimization." I find this interpretation of the ESS confusing; the definition of an ESS does not require that individual fitness be maximized (Quenette and Gerard 1992). For example, in the prisoner's dilemma game discussed below, the ESS is complete defection, but defection does not maximize individual or population fitness.

The definition of an ESS considers how a strategy may maintain itself; however, it does not consider whether this strategy may ever be established (Nowak 1990a). Genetic constraints can be a problem for ESS (Maynard Smith 1981), but one that evolution may be expected to modify and remove as discussed in the next section. The central problem is with the very fitness criterion used to define the ESS. Nowak provides an instructive example in which the payoff to fitness in a two-player game is given as an arbitrary second-order polynomial function of the two strategies (1990a). Examples are given (in terms of the coefficients of the polynomial) of the following situations: an ESS that is attainable (which we would hope to be the case), an ESS that is not attainable, and an attracting state that is not an ESS. The latter two cases underlie the dynamical insufficiency of the ESS concept. Examples of these states are also given in terms of the well-known iterated prisoner's dilemma game.

It is important to understand that these examples of qualitatively different evolutionary states all assume that the individual organism is the maximizing agent in evolution, that survival of the fittest (organism) is true, at least locally. The very definition of attainability used by Nowak assumes that evolution proceeds locally in the direction of more fit strategies. Nevertheless, an ESS may not be attainable and attracting states may not be

ESSs. This lack of robustness of game theory approaches to evolution is one of the reasons why I prefer using an explicit dynamical approach to study natural selection, especially when the very units of evolution are in the process of being created.

## MODIFICATION OF GENETIC CONSTRAINTS

Darwinian dynamics involve heritable properties of evolutionary units. There are cases where genetic constraints interfere with the ESS predictions of the outcome of natural selection (Maynard Smith 1981). In order to be more confident in employing phenotypic arguments such as the ESS, we would like to know what form genetic constraints take and what their consequences are in terms of the stable outcomes of evolution. Gayley and Michod (1990) reviewed the genetic constraints that exist in ESS models and analogous forms of frequency-dependent selection. We then used two-locus modifier models (similar to the ones studied in chapter 6 below) to determine whether long-term evolutionary modification of the genetic system can be expected to remove these constraints, leading eventually to an equilibrium predicted by the phenotypic model. We found that, in most cases, genetic constraints can be expected to be removed by the evolution of modifiers.

In simple genetic models (one locus, two alleles) of a pair of behaviors (such as cooperate versus defect studied in chapters 3, 5, and 6), there are three kinds of genetic equilibria: (i) fixation of one or the other allele, (ii) equilibration of the fitnesses of the behaviors (average fitness of cooperate equals average fitness of defect), and (iii) maximization (or minimization) of the average frequency of one of the behaviors. The third kind of equilibrium requires under- or overdominance in the expression of the behavior. Genetic constraints occer when populations get trapped at non-ESS genetic equilibria. For example, a stable but non-ESS genetic equilibrium may be encountered during the course of selection toward an ESS. Or there may simply not be enough genetic variation to reach the ESS (the ESS lies outside the range of the phenotypes). In this case the population is globally as close as possible to the ESS. However, with overdominance in the expression of the behavior there are situations in which the ESS is within the range of possible phenotypes but is not attained because of sexual recombination and segregation of less fit phenotypes.

In two-strategy games evolutionary modification of the genetic system will usually take the population to a state predicted by the phenotypic model (i.e., an ESS). Although counterexamples can be found to this in the case of two-locus modifiers, we argue that these cases are of limited interest (Gayley and Michod 1990). Therefore the theory of evolutionary modification of genetic systems validates, for the most part, commonsense phenotypic arguments. In multiple-strategy games, the situation is more complex. This is not entirely due to genetic constraints, but also to the fact that ESS is not an adequate characterization of stability even in the simplest dynamical systems that do not involve sex (Hofbauer et al. 1979; Taylor and Jonker 1978; Zeeman 1980).

## POPULATION DYNAMICS AND NATURAL SELECTION

I have been discussing how the genetic system, especially if it involves sexual recombination, can affect the character and outcome of selection. Population dynamics can also have dramatic effects on the outcome of selection (Eigen and Schuster 1979; Michod 1983b, 1984, 1986, 1991, 1995; Szathmáry 1991). The genetic system and population dynamics are related through the reproductive system and the mating system. For example, sexual reproduction makes for DNA repair and genetic mixis, but it also provides for an inherently nonlinear form of population growth. This nonlinearity stems from the simple fact that sexual reproduction requires two individuals to encounter one another, and this leads to nonlinear terms in the population dynamics involving the densities of the mates.

As we saw in the last chapter, nonlinearity in the birth process underlying population growth dramatically affects the dynamics and outcome of natural selection. When birth is a nonlinear function of density, the adaptive features of a unit of selection are no longer sufficient to predict the outcome of natural selection. "Survival of the fittest" is false, and there is no useful measure of overall adaptedness. I have introduced alternative dynamical paradigms, such as "survival of the first" and "survival of anybody," in chapters 2 and 3 to describe situations in which Darwin's conditions for natural selection are met, but the most adapted individuals do not win in selection. We discussed in chapter 3 in our consideration of population structure cases of natural selection involving spatial patterns, such as spiral waves, in which the spatial patterns and their characteristics (such as location and rotation speed for spiral waves) affect invasion and selection of the entities involved in making the spatial pattern. We also mentioned how spatial structure can profoundly affect the evolution of cooperation through its origin as a traveling wave (see figure 4-2 below). This topic is discussed in more detail below in the section "Spatial Structure and the Evolution of Cooperation." These explicit spatial models are instructive for they show that fitness (as measured by the likelihood of invasion) depends not on individual properties alone but also upon aspects of the spatial environment.

The field of adaptive dynamics is revealing new lessons about the interplay of population dynamics and natural selection (Metz et al. 1992, 1996; Rand et al. 1994; Eshel 1996; Ferriere and De Feo 1998; De Feo and Ferriere 1998; Diekmann et al. 1997; Hammerstein 1997; Geritz et al. 1997; Geritz et al. 1998). Adaptive dynamics is concerned with the rigorous classification of the dynamics of natural selection, especially with regard to the stable states of evolution and the invasion of new mutants. The individual capacities of the mutant are often not sufficient to understand invasion, which involves a complicated interplay of individual heritable capacities and the dynamical properties of the resident population. New dynamical paradigms for natural selection will likely emerge from this work.

## FITNESS MINIMA

Natural selection can lead to unfit individuals, even in the absence of genetic constraints. We may assume that evolution proceeds in a direction that increases individual fitness locally; nevertheless, negative frequency-dependent fitness effects (the cost of commonness) can reverse the outcome from what Darwin's survival of the fittest would predict. Ecological interactions such as resource exploitation, predation, and competition may trap populations at individual fitness minima.

Abrams et al. (1993) consider a class of continuous-trait models in which individual fitness (expected reproductive success) depends on the individual's trait value, $x$, as well as the average trait value in the population, $\bar{x}$ (and possibly the average value of traits in other interacting species, although for reasons of simplicity I consider only the single-species model here), so that individual fitness is given as the function $W(x, \bar{x})$. Individual trait values change in the direction that increases individual fitness, according to the fitness gradient $\partial W(x,\bar{x})/\partial x$. The fitness gradient describes the effect on individual fitness of changes in an individual's trait value. The evolution of the

73

average trait value in the population is also proportional to the fitness gradient as described in equation 4-23:

$$\frac{d\bar{x}}{dt} \propto \left.\frac{\partial W(x,\bar{x})}{\partial x}\right|_{x=\bar{x}} \tag{4-23}$$

(the average trait value is assumed to evolve on a similar time scale to the individual trait value). Evolutionarily stable equilibria require $d[d\bar{x}/dt]/d\bar{x} < 0$, which after taking the total derivative gives

$$\left.\frac{\partial^2 W(x,\bar{x})}{\partial x^2}\right|_{x=\bar{x}} + \left.\frac{\partial^2 W(x,\bar{x})}{\partial x\,\partial\bar{x}}\right|_{x=\bar{x}} < 0. \tag{4-24}$$

The interesting point about the condition for evolutionary stability given in equation 4-24 is that it does not require individual fitness to be maximized at equilibrium. Maximizing individual fitness requires just the first term on the left-hand side of equation 4-24 to be negative. Indeed, individual fitness could be at a minimum (first term positive), yet, so long as the second term is more negative than the first term is positive, equation 4-24 will be satisfied and the equilibrium will be stable. The second term of equation 4-24 is the effect on the individual fitness gradient of changing the population mean. When the second term is negative, the fitness gradient changes in the opposite direction from changes in the population mean. For the first term to be positive, individual selection must lead away from the equilibrium. For the second term to be greater in magnitude than the first term the fitness function $W(x,\bar{x})$ must be more sensitive to changes in the mean than to changes in the individual trait values. Abrams et al. (1993) give a number of ecologically interesting examples when this is the case. In the examples of fitness minima they consider, the slope of the fitness gradient changes sign as the individual trait values depart from the minimum, due to changes in the mean trait value. This reversal in sign of the fitness gradient forces the population back to the fitness minimum. The cost of commonness (negative frequency dependence) is an important factor that contributes to this result (see Abrams et al. 1993, fig. 1).

What do stable fitness minima teach us about the meaning of fitness and adaptedness under frequency-dependent selection? These explicit ecological models provide concrete examples of the general problems with frequency-dependent selection introduced earlier in this chapter. An engineering-like analysis of the organism in its environment (an environment that includes the population's average trait value) would predict that the organism is better off changing the value of the trait under consideration (in either direction). The organism is poorly adapted to its environment, because any change in its trait value should increase fitness (and it will, but only temporarily, until the other individuals in the population respond, thereby changing the mean). Consequently, we predict that natural selection (based on survival of the fittest reasoning) should change the value of the trait. Yet we know that this cannot be accomplished, for as soon as the trait changes so does the population mean, which changes the adaptive topography, returning the trait value to a fitness minimum. The take-home message is that adaptedness of the organism in its environment may have little to do with the course of natural selection when fitness is frequency dependent, even when there are no genetic constraints (see also Quenette and Gerard 1992). Again, as we first discovered in chapter 3, there is more to the dynamics of natural selection than increase of better-designed phenotypes.

## PRISONER'S DILEMMA

Cooperation drives the passage to more inclusive evolutionary units. However, the origin of cooperative interactions remains one of the most difficult problems in sociobiology. High genetic relatedness and population structure provide one kind of answer, but cooperation may exist even in situations in which genetic relatedness is low. In such situations the prisoner's dilemma game has proved to be a fruitful tool of investigation. The prisoner's dilemma game illustrates well the inherent limits of frequency-dependent selection in terms of maintaining the well-being of evolutionary units.

In this game, players have two options, to cooperate or defect. If both players cooperate, both obtain $R$ fitness units (the "reward payoff," not relatedness as in the previous section); if both defect, each receives $P$ (the "punishment payoff"); if one player cooperates and the other defects, the cooperator gets $S$ (the "sucker's payoff") while the defector gets $T$ (the "temptation payoff"). The payoff values are ranked $T > R > P > S$, and $2R > T + S$. An additive cost-benefit parameterization of these payoffs similar to equation 4-18 is sometimes useful (Brown et al. 1982). Assume that a cooperator exhibits a behavior that benefits the fitness of its partner, the recipient, by $b > 0$. The benefit is independent of the recipient's behavior. By providing its partner with the benefit $b$, the cooperator incurs a cost, $-c$ ($c > 0$). Again, this cost is independent of the recipient's behavior. An act of defection is assumed to bestow no benefit to the partner and to incur no cost to the actor. The total effect on fitness of a given interaction is assumed to be the sum of the appropriate terms; the increments to fitness are added

75

to a baseline fitness taken to be one. With this parameterization, the payoff $T$ results from receiving $b$ from the cooperator but incurring no cost: $T = 1 + b$. Similarly, one gets $R = 1 + b - c$, $P = 1$, and $S = 1 - c$. (To satisfy the constraints of the game we must have $b > c$.)

As we know, when natural selection is frequency dependent, the fitness of individuals is not maximized and may often decline. The prisoner's dilemma game illustrates this point in an especially illuminating way (Williams 1992). Selection of cooperation and defection is frequency dependent because an individuals' fitness depends upon who it interacts with. Mixed interactions between cooperators and defectors are unstable because cooperators are at a disadvantage. The best outcome for the individuals and for the population would be mutual cooperation, since that interaction receives $R$ fitness units, which is greater than $P$, the payoff for mutual defection. However, complete defection with payoff $P$ is what natural selection produces, if the game is played only once. This is because no matter how your partner behaves, defection returns a greater reward than cooperation, in spite of the fact that mutual cooperation is better than mutual defection. Even a random choice of behaviors would return a greater reward than the complete defection that natural selection produces (Williams 1992). Natural selection not only fails to maximize the fitness of individuals in the prisoner's dilemma game, it minimizes it. Matters may turn out differently, if the prisoner's dilemma game is played more than once during a generation, because repeated encounters allow for learning and modification of behavior, as occurs in the celebrated tit-for-tat strategy.

In a well-known computer tournament that simulated an iterated prisoner's dilemma (IPD for short), the simple strategy "tit-for-tat" (*TFT*) outperformed other strategies (Axelrod 1984; Axelrod and Hamilton 1981). *TFT*, which initiates a partnership by cooperating and next imitates its partner's behavior, has become a leading paradigm for cooperative behavior based on reciprocation between unrelated individuals. Further analytical work (Boyd and Lorberbaum 1987) and new computer tournaments (Nowak and Sigmund 1992, 1993; Nowak 1990b), have shown that *TFT* plays a pivotal role in an evolution of cooperation. In a situation where the unconditionally defective strategy "always-defect" (*AD*) is common, the emergence of *TFT* is the first step toward sociality. Once established, *TFT* paves the way for more robust forms of reciprocation, represented by strategies like "generous tit-for-tat" (*GTFT*), which is prone to forgive a defective act with a certain probability. In order to explain the emergence of any form of cooperative behavior, it is therefore crucial to understand how *TFT* can gain a foothold in a world of egoists, and how it can effectively serve as a stepping stone for the establishment of more generous strategies.

The problem is that a new *TFT* has no one to reciprocate cooperation. It has been argued that if the cooperators begin in small groups, they will have a chance to thrive (Axelrod and Hamilton 1981). This mechanism should apply primarily to sessile organisms (Eshel and Cavalli-Sforza 1982; Wilson et al. 1992); however, clusters of sedentary *TFT*s are easily devastated by mobile *AD* players (Dugatkin and Wilson 1991; Enquist and Leimar 1993; Houston 1993). Thus, defector mobility should preclude the emergence of *TFT*, thereby blocking the first step towards cooperation. Mobility of players threatens the well-being of cooperation for the same reason as do mixtures of genes in hypercycles: it weakens the alliances formed among cooperators (which are by their nature local) and makes it more difficult for cooperators to reap the benefits of their socializing behaviors.

The conclusion that mobility of *AD* precludes the emergence of *TFT* is based on models that assume random interactions and represent space and movement only implicitly, in terms of traveling costs. Furthermore, these models do not consider mobility of *TFT*. It seems that something very basic has been overlooked, for the very nature of an invasion event requires active movement in space or else the new type obviously cannot spread. Because players primarily meet their neighbors in space, the assumption of random interactions is unrealistic. Numerical examples (Nowak and May 1993, 1992) that include local spatial dynamics demonstrate that nonrandom interactions are important and that the usual tools of the game theory do not correctly predict the winner (Maynard Smith 1982; Metz et al. 1992).

## SPATIAL STRUCTURE AND THE EVOLUTION OF COOPERATION

Regis Ferriere and I considered a mathematical approach to iterated games that explicitly accounts for spatial dynamics and local interactions, in light of which the IPD can be reconsidered (Ferriere and Michod 1995, 1996, 1998). Our results indicate that differential mobility of *TFT* and *AD* is important for the outcome of the contest, and that there exist critical mobilities allowing for the emergence of cooperation as a traveling wave in a population dominated by *AD* (see figure 4-2 below). We have arrived at this result by studying the spatial conflict between a resident population of unconditional defectors (*AD*), invaded by a small number of cooperative players that adopt *TFT*. Individual mobility is quantified by spatial diffusion pa-

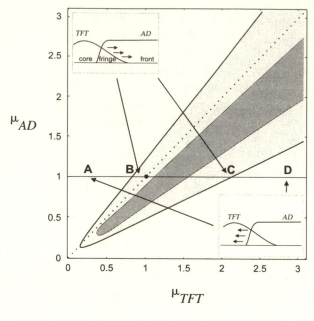

*Figure 4-2*

Increase of cooperation among mobile players in spatially structured populations. Adapted from Ferriere and Michod 1995.

rameters that may differ across strategies; *TFT*s and *AD*s move at rates $\mu_{TFT}$ and $\mu_{AD}$, respectively.

The main results are summarized in figure 4-2. Given generic values of model parameters, each curve corresponds to a particular value of the cost-benefit ratio *c/b*, and embraces the combinations ($\mu_{TFT}$, $\mu_{AD}$) for which *TFT* endowed with a mobility rate $\mu_{TFT}$ will invade a resident population of *AD*s moving at rate $\mu_{AD}$. Darker regions correspond to larger *c/b* (for clarity only two regions have been shaded). The figure shows that the defector's mobility $\mu_{AD}$ must exceed a minimum threshold for *TFT* to have a chance of invading. Insets in the figure provide schematic snapshots of the *TFT* and *AD* distributions along the spatial axis, as *TFT*'s mobility $\mu_{TFT}$ is increased while $\mu_{AD}$ is kept fixed. They illustrate the pattern of a traveling wave replacing one strategy by the other, which is predicted by the mathematical theory of traveling waves. Triple arrows indicate the direction of the wave: to the right, *TFT* invades; to the left, the *TFT* cluster is wiped out. At point *A* in figure 4-2, the value of $\mu_{TFT}$ is too low to beget invasion by *TFT*. For values slightly smaller (point *B*) or substantially larger (point $C$) than $\mu_{AD}$, TFT can increase and displace the resident $AD$ population. This is no longer possible (at point $D$) once $\mu_{TFT}$ has passed an upper threshold.

In conclusion, we find that among mobile players, a cooperative strategy can spread and take over a selfish population, even when originating from extreme rarity. Significant mobility makes the pioneering moves of cooperators towards the front of invasion less hazardous and also contributes to neutralizing those defectors that may intrude on the core of a cluster of cooperative players. In contrast to common intuition, it seems that populations that are mixed provide a good substrate for the evolution of cooperation.

Nonrandom association of behaviors, or behavioral structure, is required for cooperation to differentially reap the benefits of its socializing behavior (Michod and Sanderson 1985). In the IPD game of reciprocation, behavioral structure emerges from the learning rule inherent in TFT's strategy and from the prior experience of the players involved. In the IPD model briefly described above, space is represented explicitly and not implicitly in terms of traveling costs or predefined probabilities of interactions. Population structure is not a fixed and preordained construct but rather emerges out of the restricted and stochastic nature of individual mobility. Interactions are inherently nonrandom when viewed from the global population. In a similar way, the probability of repeated interactions, so critical to the evolution of reciprocation (Axelrod and Hamilton 1981), is not a predefined constant but is determined by the more basic parameters and variables of the model, including the individual mobilities of the strategies (Ferriere and Michod 1995, 1996, 1998).

Spatial effects are critical to reversing the direction of selection away from selfishness and towards cooperation (for another example see Boerlijst and Hogeweg 1993, 1991), and our results run counter to the conventional intuition that the increase of mobility will be bad for cooperators and good for defectors. In general, the explicit inclusion of spatial dynamics can deeply affect the outcome of evolutionary games.

In our spatial model of the IPD game, interactions occur among near neighbors who may be related by genetic descent. Low individual mobility should increase the degree of genetic relatedness between interactants, and this should facilitate the emergence of cooperation via kin selection. Our results show that a cooperative strategy can spread even in a mixed population, where the degree of kinship between players is presumably reduced—kinship appears to be unnecessary for cooperation to get started. On the other hand, very low mobility acts against the spread of cooperative behaviors in our model. This surprising result urges the development of spatial

79

models that combine both direct (game payoffs) and indirect (kin selected) effects of cooperation on fitness.

## THE PROBLEM OF FREQUENCY DEPENDENCE

New evolutionary units begin as groups of existing units. Frequency-dependent interactions among these existing units are both a source of novelty for the group and a threat to its collective well-being. Cooperative interactions are by their very nature local, and factors such as mobility and mixing provide for conflict and threaten the viability of cooperation. Likewise, in the case of interactions among cells in multicellular organisms, although cells exist in a well-defined group (the organism), replication and somatic mutation during development can lead to a divergence of cell behavior and produce conflict, which may again lead to defection. All new higher-level functions (say, at the organism level) depend upon interactions among the component cells and tissues, just as the emergent properties of a social insect colony depend upon the interactions among the individual organisms and castes that make up the colony. As we have learned, when fitness depends upon interactions, the levels in an evolutionary hierarchy exist in constant tension. The well-being of the group is under constant threat of conflict from within. Frequency-dependent selection provides little guidance for the emergence of higher-level units; instead, it is a constant threat to the viability of cooperation upon which all group-beneficial behavior depends. The temptation to defect haunts the collective well-being of the group as it leads to the tragedy of the commons (Hardin 1968), whereby any component stands to gain more by cheating than by cooperating.

Population structure and kin selection helped direct the outcome of frequency-dependent selection towards cooperation and likely aided the evolutionary passage from genes to cooperative networks of genes to cells. The cell structure served to convert the frequency-dependent tension between genes into a collective unity. We may expect similar structures and processes to protect the integrity of organisms from the threat of renegade cells. In chapters 2 and 3, we considered the early transitions in evolution from genes to gene networks to the first cell. We are now ready to consider the further increase in complexity as cells grouped to form cooperative colonies that eventually evolved into a new highly integrated individual, the multicellular organism.

# Multilevel Selection of the Organism

## A SCENARIO

To help fix ideas, let us consider a possible scenario for the initial stages of the transition from unicellular to multicellular life (Margulis 1981, 1993; Buss 1987). We may assume that reproduction and motility are two basic characteristics of the early single-celled ancestors to multicellular life, and that these single cells were likely able to differentiate into reproductive and motile states. Cell development was probably constrained by a single microtubule organizing center per cell, and, consequently, there would have been a tradeoff between reproduction and motility, with reproductive cells being unable to develop flagella for motility and motile cells being unable to develop mitotic spindles for cell division. Single cells would switch between these two states according to environmental conditions. Finally, the many advantages of large size might favor single cells coming together to form cell-groups. It is at this point that my investigations begin.

If and when single cells began forming groups, the capacity to respond to the appropriate environmental inducer and differentiate into a motile state would be costly to the cell, but beneficial for the group (assuming it was advantageous for groups to be able to move). Because having motile cells is beneficial for the group, but motile cells cannot themselves divide, or divide at a lower rate within the group, the capacity for a cell to become motile is a costly form of cooperation, or altruism. Loss of this capacity is then a form of defection, as staying reproductive all the time would be advantageous at the cell level (favored by within-group selection), but disadvantageous at the group level (disfavored by between-cell-group selection). According to this scenario (and many others), we are led to consider the fates of cooperation and defection in a multilevel selection setting during the initial phases of the transition from unicellular life to multicellular organisms.

## A MODEL FOR THE EMERGENCE OF ORGANISMS

A recent commentary on multilevel selection theory asks whether there is anything in biology that cannot be explained by individual selection acting on organisms and requires selection acting on groups (Morrell 1996). Although rhetorical, this remark reflects a dominant view in biology that most interesting questions can be addressed by viewing organisms as the sole unit of selection. But where do organisms come from? From single cells, of course. And what are multicellular organisms but groups of cells related by common descent? In this chapter, I extend a multilevel selection framework recently developed to study the evolutionary transition from single cells to multicellular organisms (Michod 1997a,b). I argue that multilevel selection theory is needed to explain the origin of the organism—that very creation which is supposed to deny the usefulness of multilevel selection in evolutionary biology. An overview of the model life cycle has already been given in figure 4-1 along with the extension of Price's covariance approach given in table 4-4.

As introduced in the section "Evolution of Multicellular Organisms" in the last chapter, organisms can be thought of as groups of cooperating cells. Selection among cells could destroy this harmony and threaten the individual integrity of the organism. For the organism to emerge as an individual, or unit of selection, ways must be found of regulating the selfish tendencies of cells while at the same time promoting their cooperative interactions. The evolution of such means of conflict regulation is discussed in the next chapter. The purpose of the present chapter is to use multilevel selection theory to study the levels of within-organism variation that may arise by mutation and selection during development, and to explore the consequences of this variation for the levels of cooperation and fitness attained in the cell-group or adult form. More generally, my goal is to develop a theoretical framework to study the emergence of individuality and new levels of fitness. In the present chapter, I often use the word "organism" to refer to cell groups, even though there are no truly organism functions. The topic of individuality is considered further in the next chapter.

Multicellular organisms often develop from a zygote, and the replication of cells during development is indicated by the solid vertical arrows in figure 4-1. In this case, all cells are clonally derived from a single zygote and are related genetically. This case provides a kind of "worst case" for the consequences of within-organism variation. If the organism were a mixture of cells of different ancestries, as is likely the case for a migrating slug in the cellular slime mold *Dictyostelium discoideum,* this could only increase the levels of within-organism variation and change.

During the proliferation of cells throughout the course of development, deleterious mutation can lead to the loss of cooperative cell functions (such as the ability to become motile in the scenario above). I represent cell function in terms of a single cooperative strategy, as in the previous chapters.

Because deleterious mutation leads to loss of cell function, it is assumed to produce defecting cells from cooperative cells. Deleterious mutation could also produce completely defective cells with no capacity to replicate or survive. Such mutations are disadvantageous at both the cell level and the level of the group. They too can select for modifiers of within-organism change (Michod and Roze 1998). I ignore in the model mutations that are beneficial at both the cell and group levels, because they will obviously spread in the population. Mutant cells no longer take time and resources to cooperate with other cells and as a result may replicate faster or survive better than cooperating cells. Because of mutation and different rates of replication and survival of the different cell types, gene and genotype frequencies change during development of the organism (as represented by $\Delta q_j$ in figure 4-1). After development, the adult organism contributes gametes to start the next generation (dashed arrows in figure 4-1) according to its group or organism fitness. As discussed below, adult fitness is assumed to depend on adult size (number of cells) and functionality (level of cooperation among cells). In the case of sexual reproduction these gametes fuse randomly to form a diploid zygote. In the case of asexual reproduction the gametes become zygotes and develop directly into the adults of the next generation. Gene and genotype frequencies change in the population of organisms due to differences in fitness (gamete output) between the adult organisms. The two components of frequency change—within organisms and between organisms—give rise to the total change in gene frequency, $\Delta q$, in the population.

Cooperation trades fitness between levels in a selection hierarchy (table 5-1). In the case of cells within organisms, cooperation benefits the organism while detracting from the fitness of cells. Because it takes time and energy to help the group, cooperating cells replicate more slowly or survive worse than mutant defecting cells. I also assume that adult size is indeterminate and that organism fitness is a function of both adult size and the level of cooperativity among the organism's cells. The size of the group (the organism) depends upon the rate at which the cells divide. As a result of these assumptions, organisms composed of many cooperating cells end up being smaller but more functional than organisms made up of many mutant defecting cells. Because organism fitness is assumed to be a function of both size and functionality (as measured by the level of cooperation among the organism's component cells), cooperation among cells has both positive and negative effects on group (that is, organism) fitness. Even in the absence of within-organism variation (created by within-organism mutation and selection), for cooperation to exist at all, the benefit to the organism of increased

83

| | Level of Selection | |
| Cell Behavior | Cell | Group (organism) |
| --- | --- | --- |
| Defection | (+) Replicate faster or survice better | (+) Larger (−) Less functional |
| Cooperation | (−) Replicate slowly or survive worse | (−) Smaller (+) More function |

*Table 5-1*

Effect of cooperation on fitness at cell and organism level. It is assumed that growth is indeterminate and the sizes of the adults vary depending upon the composition of cells. The +/−symbol means positive or negative effects on fitness at the cell or organism level. Compare with table 3-1.

cooperation among its constituent cells must overcome the cost to the organism of a smaller adult size.

Fixed adult size for all genotypes removes one of the temptations to defection at the cell level, since there would be no effect on group (organism) size of a defecting cell's faster rate of replication. There would still be an advantage to defection during development because defecting cells replicate faster, it is just that the overall size of the organism would not be affected. Viewed in this way, fixed organism size may be seen as an adaptation to help maintain the integrity of organisms by removing one of the costs of cooperation. Indeed, the conditions for the evolution of cooperation are more relaxed when adult size is fixed (J. Lie and R. E. Michod, unpublished data).

## RECURRENCE EQUATIONS

I assume that a single locus with two alleles, cooperate $C$ and defect $D$, determines the way cells behave and interact. The definition of terms and variables is given in table 5-2. I refer to the organisms in terms of the genotype of the zygote at the cooperate defect locus. Because of within-organism change during development, the adult stage may have cells with different genotypes than that of the zygote. The $k$ variables in table 5-2 refer to the numbers of cells of different genotypes in the adult stage. There are two forces that may change gene frequency between the zygote and adult stages and determine the values of the $k$ variables: mutation during cell replication and cellular selection. Mutation is assumed to lead to loss of cooperation and tissue function. Mutation also increases the variance among cells and enhances the scope for selection and conflict among cells within organisms. Cellular selection is represented in table 5-3 below as differences in cell replication rate or the probability of cell survival.

The fitness of the adult form, represented by $W_j$, is the absolute number of gametes produced, and is assumed to depend upon both the number of cells in the adult and how the cells interact. Cooperation among cells increases the fitness of the adult (parameter $\beta$) but noncooperating cells replicate faster (parameter $b$ in table 5-3) and produce a larger but less functional adult (assuming indeterminate adult size). The simple model of cooperation considered in table 5-2 assumes a linear dependence of adult fitness on the frequency of cooperating cells, although more complex models could easily be incorporated into the framework considered there. Organism size is assumed to be indeterminate and to depend on the time available for development as well as the rate at which cells divide. Indeterminate growth is most applicable to organisms like plants or clonal invertebrates in

| Term | Meaning |
|------|---------|
| $k_{ij}$ | Number of $i$ cells in the adult stage of a $j$-zygote: $i, j = C, D$ |
| $k_j$ | Total number of cells in adult stage of $j$-zygote after development: $j = C, D$ |
| $W_D$ | Individual fitness of $D$-zygote: $W_D \propto k_D$ |
| $W_C$ | Individual fitness of $C$-zygote: $W_C \propto k_{CC} + k_{DC} + \beta\, k_{CC}$ |
| $\beta$ | Benefit to adult organism of cooperation among its cells |
| $q_j, \Delta q_j$ | Initial frequency, and change in frequency, of $C$ gene within $j$ organisms |
| $q, \Delta q$ | Initial frequency, and change in frequency, of $C$ gene in total population |
| $h_W^2$ | Heritability of fitness at the cell-group or organism level, defined as $\mathrm{Cov}(W_P, W_O)/\mathrm{Var}(W_P)$, where $W_P$ and $W_O$ are the fitness of parents and the average fitness of offspring, respectively, for the two types; in haploids, $h_W^2 = k_{CC}/k_C$ |

*Table 5-2*

Multilevel selection model of the organism. The proportionality constant for $W_C$ and $W_D$ is assumed to be the same and can be ignored as it does not affect any calculations. The heritablility given for haploids is at equilibrium (equation 5-4 below).

85

which multicellularity likely first evolved. As already mentioned, indeterminate growth makes matters more difficult for organisms, because it allows selfish cells to reap additional fitness benefits by virtue of larger adult size.

With these definitions it is straightforward to write down the new gene frequency in the next generation in equation 5-1. For simplicity of presentation, I focus on haploidy in the text. Extensions of the model to diploidy are discussed in appendix A.

$$q' = \frac{qW_C\left(\dfrac{k_{CC}}{k_C}\right)}{\overline{W}} = \frac{qW_C h_W^2}{\overline{W}},$$ (5-1)

$$\overline{W} = qW_C + (1-q)W_D.$$

The bracketed term is the frequency of $C$ alleles in the cells of a $C$ adult and appropriately weights the total gametic output to consider only those gametes containing $C$ alleles. This bracketed term equals the heritability of fitness defined as the regression of average offspring fitness on adult fitness (see figure 6-5 below). Because of within-organism change, the heritability of fitness is not unity, as it should be for "proper" organisms under asexual haploidy when there is no environmental variance (as it is for $D$ zygotes in the model).

## WITHIN-ORGANISM MUTATION SELECTION MODEL

As cells proliferate within the developing organism, mutations occur, leading to loss of tissue function and cooperativity among cells (rate $\mu$ per cell division). I consider only mutations from $C$ to $D$ (no back mutation) as this represents a worst case for the evolution of intercellular cooperation. Ignoring back mutation is reasonable, because it is far easier to lose a complex trait like cooperativity among cells than it is to gain it.

A simple model of mutation and cellular selection as represented in table 5-3 extends previous work (Otto and Orive 1995; Michod 1997a) to include survival selection among cells in addition to replication. As explained further in figure 5-1, this model gives $k_{CC}$, $k_{DC}$, $k_D$, $W_C$, and $W_D$ in terms of the parameters of mutation, selection, and the time for development. The time for development, $t$, is measured on the scale of time taken for a cooperating cell to divide (without loss of generality, I take $c = 1$) and so $t$ gives

| Term | Meaning |
|------|---------|
| $\mu$ | Within-organism mutation rate from $C$ to $D$ per cell division |
| $t$ | Time for development |
| $c$ | Rate of cell division for cooperating cells ($c = 1$ in most analyses) |
| $b$ | Advantage to cell of defection (in terms of replication rate; $b > 1$) |
| $cb$ | Rate of cell division for defecting cells |
| $s_C, s_D$ | Probability of cell death for $C$ and $D$ cells, respectively |

$$k_{CC} = \left[2s_C(1-\mu)\right]^{ct}$$

$$k_{DC} = \sum_{x=1}^{ct} 2\left[2s_C(1-\mu)\right]^{x-1} \mu s_D \left[2s_D\right]^{b(ct-x)} \text{ ; see figure 5-1}$$

$$W_D = \left(2s_D\right)^{bct}$$

$$W_C = k_{CC} + k_{DC} + \beta\, k_{CC}$$

*Table 5-3*

Within-organism mutation selection model.

the number of cell divisions per individual generation assuming no selection. Crude estimates of adult size may be obtained by raising 2 to the development time, $2^t$. Selection at the cell level, resulting from differences in the rate of cell division or cell survival, means that the actual number of cells in the adult stage will be different.

## MUTATION RATE

Although the model considers a single locus, there are likely to be many loci that affect tissue function and cooperativity among cells. For this reason, I like to think of this single locus as representing the cumulative effect of deleterious mutation at all loci leading to loss of cooperation. This is not rigorous modeling but provides a better understanding of what I would like the model to represent. The mutation rate parameter in the model does not pertain to cells in modern multicellular organisms but rather to cells on the brink of coloniality during the period in which multicellular life first arose. Indeed, the model predicts that one of the ways of coping with within-organism change is to select for modifiers that lower the mutation rate. Consequently, the mutation rate in modern organisms likely results from the very processes being modeled here. For these reasons, below I consider mu-

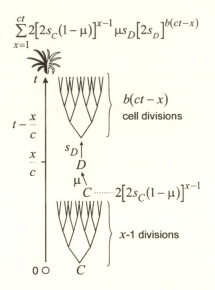

$$\sum_{x=1}^{ct} 2\left[2s_C(1-\mu)\right]^{x-1} \mu s_D \left[2s_D\right]^{b(ct-x)}$$

*Figure 5-1*

Calculation of the number of defect cells (D) in the adult stage of a cooperate (C) zygote, $k_{DC}$. The figure explains the formula given in table 5-3. For any single cell generation I assume replication, mutation, and then survival. A C cell divides for $x - 1$ divisions and then mutates to D during cell division $x$. Assume there are $N_x$ cells in generation $x$. After cell division there are $2N_x$ cells and these survive with probability $s_C$ or $s_D$, for C and D cells, respectively. The total number of D cells in the adult organism is represented by the color black in the adult form. Consider C cells that have divided, survived, and not yet mutated for $x - 1$ divisions, and are now completing cell division, $x$. The total number of these cells is $2[2s_C(1 - \mu)]^{x-1}$. Some of these cells ($\mu$) will have mutated for the first time and the resulting mutants will survive to the next cell division with probability $s_D$ and then undergo $b(ct - x)$ more cell divisions. The time taken to get $x$ cell divisions is $x/c$. The time left is $t - x/c$. The number of cell divisions the mutant will undergo is then $cb(t - x/c) = b(ct - x)$. The sum is over all possible cell divisions $x$. The model has been extended to propagule reproduction, where the adult fragments into offspring groups (Michod and Roze 1998).

tation rates that are high when viewed as pertaining to a single locus in a modern organism.

Modern microbes as diverse as viruses, yeast, bacteria, and filamentous fungi, have a genome-wide deleterious mutation rate of 0.003 per cell division even though they have widely different genome sizes and mutation rates per base pair (Drake 1991). For independently acting mutations, a mutation rate of 0.003 yields a low value for the mutation load; the average fitness is approximately $e^{-0.003} \approx 0.997$ (Haldane 1937; Hopf et al. 1988). The genome-wide mutation rate in modern multicellular organisms is much higher than in microbes, for example; *Drosophila* has a genome-wide rate (expressed on a haploid basis) of approximately 0.5. Of course, the "organism" I have in mind here is nowhere near as sophisticated as *Drosophila*, being on the threshold of the transition from unicellular to multicellular life.

The mutation rate in modern microbes of $\mu = 0.003$ has been attained through the evolution of modifiers over billions of years; these modifiers presumably balanced the benefits of reducing the mutation rate with the physiological costs of doing so. For this reason, I think a much higher overall deleterious mutation rate likely held for the unicellular progenitors of multicellular life. On the other hand, not all deleterious mutations are disadvantageous at the organism level and advantageous or neutral at the cell level, which are the cases I primarily study below. Some mutations, probably many, will be disadvantageous at both levels. Mutations that are deleterious at both levels also select for a germ line and other mediators of within-organism change (Michod and Roze 1998). In any event, even though there are good reasons to expect the mutation parameter in the model to be both higher and lower, I use $\mu = 0.003$ as a point of departure in many of my studies reported in the next chapter.

## COVARIANCE METHODS

The recurrence equation 5-1 above is derived by directly monitoring the numbers and frequencies of cells at the different life stages. An alternative method for representing selection in hierarchically structured populations is Price's covariance approach (Price 1995, 1972a, 1970), introduced in the previous chapter (table 4-4). Although the covariance approach gives the same change in gene and genotype frequencies as the direct methods, it will help us to better understand the results. Price's approach posits a hierarchical structure in which there are two selection levels; in our case, between cells within organisms—viewed as a group of cells—and between organisms within populations. As discussed in the last chapter, both levels of selection can be described by a single equation, which in the present situation becomes

$$\Delta q = \frac{\text{Cov}_q[W_i, q_i]}{\overline{W}} + E_{Wq}[\Delta q_i], \qquad (5\text{-}2)$$

89

with the vectors $\mathbf{q} = (1 - q, q)$, and $\mathbf{Wq} = (W_D(1 - q), W_c q)$ used as weights. Variables $q$ and $q_i$ are the frequencies of a gene of interest in the total population and within zygotes; $\mathrm{Cov}_q[x, y]$ and $\mathrm{E}_{\mathbf{Wq}}[x]$ indicate the weighted covariance and expected value functions, respectively.

The Price equation 5-2 gives the same change in gene frequency as the direct method given in equation 5-1, but the covariance approach is more illuminating and is well suited to studying conflict and cooperation among cells within organisms. Mutation and conflict among cells results in within-organism change as represented by the within-group component of equation 5-2, $\mathrm{E}_{\mathbf{Wq}}[\Delta q_i]$. The fitness of the emerging higher-level unit may be represented by the first component of equation 5-2, $\mathrm{Cov}_q[W_i, q_i]/\overline{W}$. The heritability of fitness is defined as the regression of offspring fitness on parent fitness (table 5-2).

## THE RISK OF DEVELOPMENT

Along with the many advantages of larger organism size (more kinds of cell interactions, bigger organisms eat smaller ones) come the risks inherent in the greater opportunity for within-organism change. One component of the total change in the population occurs within organisms during development (recall the second term on the right-hand side of the Price equation 5-2). The buildup of variation during development has been studied previously (Michod 1997a,b) and can be a strong force favoring defection. Recall that cooperation is assumed to take time away from cell division (or to decrease cell survival). Within-organism variation during development increases as a function of development time, $t$, mutation rate, $\mu$, and within-organism selection, $b$. In addition, the parameters interact, especially selection and mutation. While being an essential ingredient, the mutation rate alone is not the critical force. Rather it is both mutation and selection (as determined by the development time and replication and survival at the cell level) that determine the amount of within-organism change. More time for development means not only a larger adult but also greater variation in the adult form. As the variation within organisms increases, the frequency of the $C$ allele within organisms must decrease because it is at a disadvantage at the cellular level. For simplicity, I assume here that within-organism selection results only from differences in cell replication rate; I ignore cell death by setting $s_C = s_D = 1$. Cell death results in additional selection, thereby increasing the level of within-organism variation; in addition, because of

90

cell death, more cell divisions are required to attain a given adult size, again increasing the level of within-organism change.

## INCREASE OF COOPERATION

For an organism to emerge as a new evolutionary unit, greater levels of harmony and cooperation must be attained among cells. Under what conditions can cooperation increase among cells within the emerging organism? This question is answered by studying the conditions under which the fixation equilibrium of complete defection is unstable. In the case of haploidy the eigenvalue is given in equation 5-3:

$$\lambda = \frac{W_C}{W_D} \frac{k_{CC}}{k_C} = \frac{W_C}{W_D} h_W^2. \qquad (5\text{-}3)$$

This eigenvalue is a product of two components. The first is a ratio of organism fitnesses and is identical to the standard condition for increase based on between-organism or individual selection when there is no within-organism variation and change. The second component is a diluting effect resulting from within-organism change and equals the fraction of the cells in the adult organism that contain the $C$ allele. The second component tends to unity as within-organism change, as determined by the mutation rate and within-individual selection, tends to zero. The second component equals the heritability of fitness (at equilibrium) $h_W^2$, defined as the regression of offspring fitness on adult fitness (table 5-2 and equation 5-6 below).

Increase of the $C$ allele when rare requires instability of the $D$ fixation equilibrium and this occurs when the eigenvalue given in equation 5-3 is greater than 1, $\lambda > 1$. The conditions for increase of cooperation are given in figure 5-2 as a function of the benefit of cooperation to the organism ($\beta$), the mutation rate ($\mu$), and time of development ($t$) for a fixed value of 5% selection at the cell level, $b = 1.05$. Cooperation increases for parameter values above the surface in panel A in figure 5-2. The surfaces in panels B–D are two-dimensional slices through the three-dimensional surface in panel A. In panel B, critical values of $\beta$ and $t$ are plotted for fixed values of $\mu = 0.001$ and $0.0001$. Parameter values above the surfaces in panel B allow cooperation to increase. There is little difference in the curves for the two mutation rates, which differ by an order of magnitude. However, the increase of cooperation is quite sensitive to changes in $\beta$ and $t$. A similar con-

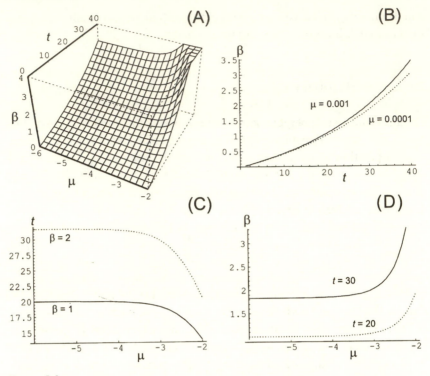

*Figure 5-2*

Increase of cooperation from rarity. Conditions for $\lambda > 1$ (equation 5-3) as a function of the time for development, $t$, the mutation rate, $\mu$, and the benefit of cooperation, $\beta$, assuming the replication advantage for defection $b = 1.05$ and $s_C = s_D = c = 1$. As explained in the text in regard to equation 5-3, the region for increase of cooperation is identical to the region for existence of a stable internal equilibrium in the frequency of the cooperation allele. In the three-dimensional panel A, parameter values above the surface allow the increase of cooperation. Panels B–D, are two-dimensional slices through the three-dimensional surface in panel A. Parameter values above the surfaces in panels B and D, and below the surface in panel C, allow increase of cooperation. In panel B, the mutation rate is fixed at values $\mu = 10^{-3}$ and $10^{-4}$. In panel C, the benefit of cooperation is fixed at values $\beta = 1$ and 2. In panel D, the development time is fixed at values $t = 20$ and 30. See text for more discussion. Adapted from Michod 1997b.

clusion is reached in the other panels. In panel C, critical values of $t$ and $\mu$ are plotted for fixed $\beta = 1$ and 2. Parameter values below the curves in panel C allow cooperation to increase. Over a wide range of mutation rates the curves are relatively flat, indicating again little effect of the mutation rate until it reaches values of approximately $\mu = 0.001$. By doubling the benefits of cooperation from 1 to 2, organisms may increase their adult body size by approximately 50% (from $t = 20$ to $t = 30$) over a wide range of mutation rates. Body size scales geometrically by the factor $2s$ with time for development, $t$. With $t = 30$ time units for development there will be approximately $10^9$ cells in the adult form (assuming no cell death). Similar conclusions may be reached from panel D. The replication advantage, $b$, and the time for development, $t$, have similar effects (results not shown for reasons of space). An advantage in cell replication, increased $b$, is compounded during development.

The results of figure 5-2 may be understood in terms of the differing opportunities for within- and between-organism selection. Indeed, the critical surfaces graphed in figure 5-2 can be obtained by an alternate approach of requiring the between-organism component of equation 5-2 to be greater than the within-organism component: $\text{Cov}[W_i, q_i]/\overline{W} > \text{E}[\Delta q_i]$ (for small $q$). Consequently, cooperation and harmony among cells may increase when the heritable covariance of fitness at the organism level overpowers the within-organism change toward defection.

## LEVEL OF COOPERATION AMONG CELLS WITHIN ORGANISMS

An organism composed of completely cooperating cells cannot occur, because of recurrent mutation, $\mu > 0$, leading to loss of cell function. The best we can hope for is that cooperation among cells increases when rare and reaches high levels within the organism. The internal equilibria of the system are of interest for this reason. For haploidy there is a single possible internal equilibrium given in equation 5-4 (it requires $W_C > W_D$ to be meaningful):

$$\hat{q} = \frac{W_C h_W^2 - W_D}{(W_C - W_D)}. \tag{5-4}$$

Again, we see the diluting effect of heritability in weighting the fitness of $C$ organisms. The eigenvalue describing the stability of this equilibrium, $\hat{q}$, is the inverse of the eigenvalue at $\hat{q} = 0$ or $1/\lambda$, where $\lambda$ is given by equation 5-3. When cooperation increases from rarity it reaches a stable internal equilibrium. Consequently, the regions for increase of cooperation given in figure 5-2 are also regions for the existence and stability of biologically meaningful equilibria.

The level of cooperation and synergism attained among cells in organisms is studied in figure 5-3 (using equation 5-4) as a function of the development time in panel A, the mutation rate in panel B, the benefit to organisms of cooperation in panel C, and the advantage to cells of defecting in panel D. Many combinations of parameter values have been studied; however, figure 5-3 is typical, especially with respect to the qualitative threshold nature of the curves. One interesting aspect of the curves is that cooperation typically remains high up to a limiting value for the parameter considered, even though within-organism variation and change is increasing continuously as the parameters $t$, $\mu$, and $b$ increase (Michod 1997a,b). This is in contrast to average organism fitness, which tends to scale with the parameters considered (figure 5-4). I will consider this matter further below when discussing that figure.

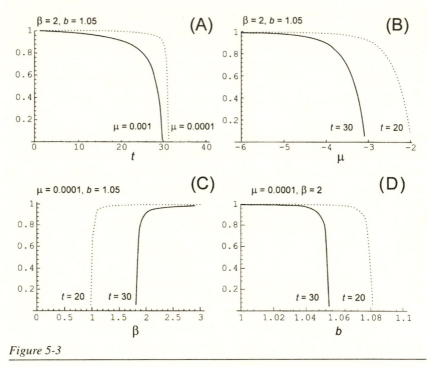

*Figure 5-3*

Level of cooperation. The vertical axis in all panels is the equilibrium frequency of cooperation, $\hat{q}$ (equation 5-4), as a function of development time, $t$ (A), mutation rate, $\mu$ (B), benefit of cooperation, $\beta$ (C), and benefit of defection, $b$ (D). Fixed parameter values are given above each panel. The dotted line in all panels corresponds to the lower level of within-organism change ($\mu = 0.0001$ in panel A and $t = 20$ in panels B–D. Only locally stable equilibria are shown. Figure adapted from Michod 1997b.

In panel A of figure 5-3 we see that there is a rather abrupt limit in size for haploid organisms (at about $10^9$ cells for $t = 30$). Cooperation remains high up to this threshold and then drops off precipitously in an almost step-like manner. The weaker the mutation rate the more precipitous the drop (compare the two curves in panel A); however, the mutation rate does not seem to drastically affect the limit value in development time, which may be viewed as a limit value in adult size. Such a threshold always exists but the exact value of the threshold depends on the other parameters, especially $b$. There are similar threshold values for the selection parameters at the two levels, as can be seen in panels C and D. The steplike nature of the level of cooperation suggests that when multicellular haploid organisms can exist they attain a high level of cooperation and synergism among their component cells. Poorly organized and barely functional haploid creatures are not predicted by these results. More intense selection at the cell level (higher levels of $b$) does not qualitatively affect the lessons drawn from panels A–C in figure 5-3. The general shape of the curves is the same but they are offset to the left in all panels except in panel C where the curves move to the right. In other words, for larger $b$, the thresholds mentioned with regard to figure 5-3 occur for smaller values of $t$ (smaller organisms) and $\mu$ and greater values of $\beta$.

## Fitness of Organisms

To further clarify the forces and factors shaping the evolution of cooperation among cells within organisms, I now consider organism fitness, its population average, heritability, and covariance with the genotype of the zygote, as the parameters describing selection and conflict change.

Average organism fitness at equilibrium is $\hat{\bar{W}} = \hat{q}W_C + (1 - \hat{q})W_D$. Using equation 5-4 and simplifying, we get the average fitness at equilibrium in the form

$$\hat{\bar{W}} = h_W^2 W_C,$$ (5-5)

where as before (table 5-2) the heritability of organism fitness is

$$h_W^2 = \frac{k_{CC}}{k_C}.$$ (5-6)

Average fitness at equilibrium (equation 5-5) depends on the parameters describing within-organism change and selection ($t$, $b$, $\beta$, and $\mu$) through their

95

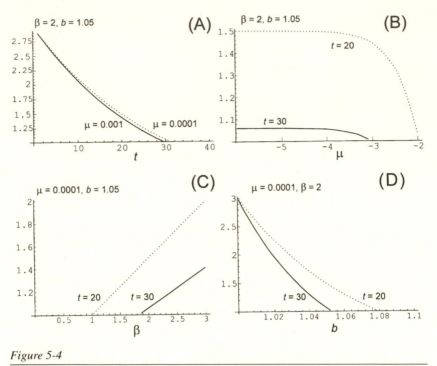

*Figure 5-4*

Average relative organism fitness at equilibrium, $\hat{w}$. See text for definition of $\hat{w}$. The panels and parameter values correspond to those given in figure 5-3. Statistics are based on relative organism fitness obtained by dividing absolute fitness by the absolute fitness of the defecting genotype, *D,* so that the relative fitness of *D* zygotes is unity at all points in all panels. Note that organism fitness does not depend on the gene and genotype frequencies in the population of organisms (it does depend on the frequencies of cell types within the organism, however). The legend is the same as in figure 5-3. Adapted from Michod 1997b.

effects on heritability and absolute fitness. For the purpose of studying average fitness at the cell-group or organism level, I use relative fitness in figure 5-4. This is because $C$ and $D$ zygotes grow to very different sizes depending on the parameters of within-organism change, and it is easier to compare their fitness by standardizing to the size (absolute fitness) of $D$ organisms. Relative fitness at the organism level is obtained by dividing absolute fitness by the absolute fitness of $D$ organisms, so that $w_C = W_C/W_D$ and $w_D = W_D/W_D = 1$. At equilibrium, we have average relative fitness given by $\hat{w} = h_W^2 w_C$, instead of equation 5-5.

In figure 5-4, the average relative fitness of organisms is graphed for the equilibrium populations studied in figure 5-3. Average fitness (relative to $D$ organisms) declines with the parameters describing within-organism variation and selection: development time $t$ is shown in panel A, mutation rate $\mu$ in panel B, and the advantage of defection to cells, $b$, in panel D. Average relative fitness increases with the benefit of cellular cooperation, $\beta$, as shown in panel C. The effect of the mutation rate $\mu$ on average fitness is small until high mutation rates are reached, as shown in panel B.

The model assumes that absolute fitness increases with adult size, and so absolute fitness always increases with development time, $t$, and the replication advantage of $D$ cells, $b$, because more time for development or faster cell division means larger organisms. This is true for $C$ organisms, even though the level of cooperation and functionality of the $C$ organism declines with development time or the replication advantage of defecting cells (because there is more opportunity for defecting cells to spread). Functionality of the adult is assumed to increase linearly with the number of cooperative cells in the adult stage according to parameter $\beta$ (table 5-2). Admittedly, the assumption of linearity with regard to the benefits of cooperation is simplistic; its main virtue is that it permits mathematical analysis and that it represents a kind of "worst case" for the spread of cooperation. In any event, the loss of functionality in $C$ organisms as development time increases (or as defecting cells replicate faster) is swamped by the geometric nature of cell division, leading to larger adult size and larger absolute fitness. Although absolute fitness increases with development time, $t$, or the replication advantage of defection, $b$, the population does worse (as panels A and D in figure 5-4 show) relative to a situation with no cooperation (as is the case in $D$ organisms). The population is doing worse as conflict increases, because $C$ organisms are doing worse (relative to a situation with no cooperation). This represents the risk of development (discussed above) for the spread of cooperation and the emergence of individuality at the new cell-group or organism level.

Apart from the predictable relations shown in figure 5-4, the average fitness of organisms is not very informative for explaining the evolution of cooperation. Average fitness varies only slightly in the regions of differences observed in figure 5-3 and does not explain many important aspects of the curves, especially the steplike nature of panels A, C, and D. For example, as already mentioned, cooperation remains high up to the limiting development time of about $t = 30$ (panel A of figure 5-3) even though the average relative fitness of the population declines steadily over this region (panel A of figure 5-4). Similar differences may be observed in panels D of figures 5-3 and 5-4 for the increases in the advantage of defection, $b$. The average organism fitness declines continuously as within-organism change builds up with increasing $b$ and $t$ (see figure 5-5 below), but the level of cooperation within organisms is buffered from this change. After considering figure 5-5 below, I discuss the underlying reasons for this buffering effect (see also appendix A). In contrast to the differing responses of cooperation and average individual fitness to changes in the parameters $t$ and $b$, the equilibrium levels of cooperation and the average individual fitnesses behave in an almost parallel fashion, as the mutation rate $\mu$ increases (compare the similar nature of panels B of figures 5-3 and 5-4).

The different components of Price's equation 5-2—the variances, covariances, and regressions involving organism fitness, within-organism selection, and individual gene frequency—all have something different to tell about the underlying causes of evolutionary change (see also appendix A). In figure 5-5, I consider the first part of the Price equation 5-2, $\mathrm{Cov}_q[W_i, q_i]/\overline{W}$, the weighted covariance of individual fitness with individual frequency, for the equilibrium populations studied in figure 5-3. In appendix A (equation A-1 and figure A-1), I show that in haploid populations the regression of individual fitness on individual gene frequency is independent of population gene frequency and equals

$$\mathrm{Reg}[W_i, q_i] = W_C - W_D. \tag{5-7}$$

The covariance is more complicated because it involves the population variance in individual fitness and individual frequency, and so depends on population gene frequency. As discussed in appendix A, it equals $\mathrm{Cov}_q[W_i, q_i] = q(1 - q)(W_C - W_D)$. In equilibrium populations, using equations 5-4 and 5-5, we obtain the first component of the Price equation in the form

$$\frac{\mathrm{Cov}_{\hat{q}}[W_i, q_i]}{\overline{W}} = \frac{1 - h_W^2}{h_W^2} \frac{(W_C h_W^2 - W_D)}{(W_C - W_D)}, \tag{5-8}$$

where heritability is given by equation (5-6). Equation 5-8 is plotted in figure 5-5. The panels and parameter values in figure 5-5 correspond to those in figure 5-3.

Populations at equilibrium must exactly balance the two levels of selection—within and between organisms—because within-organism change is never zero, due to mutation. Higher values of fitness covariance imply correspondingly high levels of within-organism change, or else the population would not be in equilibrium. For this reason, the weighted covariances

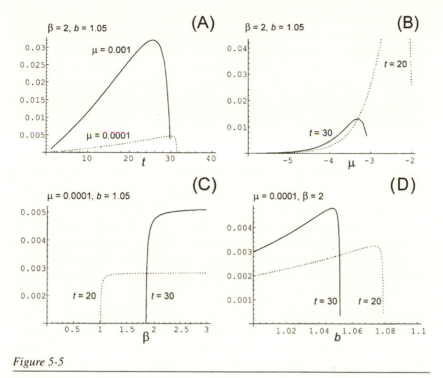

*Figure 5-5*

Fitness covariance at equilibrium. The vertical axis in all panels is the first part of the right-hand side of the Price equation: $\mathrm{Cov}_q[W_i,q_i]/\overline{W}$ (equation 5-2). At equilibrium this must equal in magnitude the weighted average of within-organism change, $-\mathrm{E}_{w_q}[\Delta q_i]$ (equation 5-2). This is because the populations are at equilibrium and so equation 5-2 must equal zero. Therefore the two parts on the right-hand side of equation 5-2 must equal one another in magnitude (the first part is positive and the second part is negative). Statistics are based on relative organism fitness as discussed in the legend to figure 5-4. The panels and parameter values correspond to those given in figure 5-3. The rest of the legend is the same as in figure 5-3. Adapted from Michod 1997b.

graphed in figure 5-5 must equal the magnitude of the average within-organism change—that is, the negative of the second part of the right-hand side of equation 5-2. Consequently, the curves in figure 5-5 may be interpreted in two ways: either as the weighted covariance of organism fitness with individual frequency or as the amount of within-organism change (within-organism change is negative because the $C$ allele is disfavored by

99

within-organism selection). As populations reach the limits at which selection can no longer maintain cooperation (discussed in reference to figure 5-3), within-organism change increases until it overwhelms selection at the organism level (figure 5-5, panels A, B, and D).

I have already pointed out that the dynamics of change are relatively insensitive to the mutation rate until rather high levels of mutation are reached. There is little effect of the mutation rate $\mu$ on individual fitness and the equilibrium level of cooperation over several orders of magnitude until the limiting values are reached (compare the similar nature of panels B of figures 5-3 and 5-4). Likewise, there was little effect of the mutation rate on the conditions for increase of cooperation in the first place until the mutation rate became high (recall panels C and D of figure 5-2).

The level of cooperation attained in the multilevel setting studied here is relatively insensitive to this buildup of within-organism change, until the limiting values of the parameters are reached (figure 5-3). This is consistently the case for the mutation rate as has already been discussed; average fitness and the increase and level of cooperation are all insensitive to the mutation rate until the limiting values of the parameter are reached. However, for the other parameters of within-organism change ($b$ and $t$), this is rather surprising, because individual fitness of the organism is quite sensitive to increases in these parameters (panels A and D of figure 5-4), yet the level of cooperation remains high until the parameter limits are reached (figure 5-3). The decline in organism fitness with increasing $b$ and $t$ is compensated by increasing the covariance of organism fitness with zygote gene frequency, so that the existing level of cooperation and harmony within the organism is preserved (figure 5-5), at least up to the parameter limits.

With regard to panel C of figure 5-5, there is little effect of changes in $\beta$ on the amount of within-organism change. This is because changes in the selection parameter at the organism level, $\beta$, cannot affect within-organism change and so the curves in panel C should be relatively flat until the threshold is reached.[1] As the threshold is reached, small changes in $\beta$ drastically change the weighted covariance and weighted average level of within-organism change. The fitness of the organism level is sensitive to changes in the benefit of cooperation, as panel C of figure 5-4 shows. The dynamics of increase of cooperation genes are sensitive to these effects on individual fitness (as figure 5-2 shows), even though the level of cooperation ultimately attained in the population is not sensitive to the effect of cooperation on individual fitness (as shown in panel C of figure 5-3).

## EFFECT OF SEX AND DIPLOIDY ON THE EMERGING ORGANISM

Sex and diploidy are studied in detail in appendix A and are shown to affect the level of conflict and variation within the emerging organism in profound ways (also see Michod 1997a). Sex helps diploids maintain higher heritability of fitness under more challenging conditions especially when there is great opportunity for within-organism variation and selection. With sex, as the mutation rate increases, and concomitantly the amount of within-organism change, more and more of the variance in fitness is heritable, explained by genes carried in the zygote, regardless of whether the mutations are additive (figure A-2) or recessive (figure A-6) and regardless of whether the equilibrium involves heterozygote superiority or not (panels D or C, respectively, of figures A-2 and A-6). Sex allows the integration of the genotypic covariances in a way not possible in asexual populations.

The increase in complexity during the evolution of multicellularity required new gene functions. However, the increase in genome size led to an increase in the deleterious mutation rate. I pointed out earlier in this chapter how vastly different (over two orders of magnitude) are the genome-wide mutation rates in modern multicellular organisms ($\approx 0.5$ on a haploid genome basis) and modern microbes ($\approx 0.003$). Once multicellularity evolved, the continued evolution of multicellular organisms required new gene functions with a corresponding increase in genome size. With an increasing genome size came the problem of increasing rates of deleterious mutation. It is often noticed that diploidy helps multicellular organisms tolerate this increase in mutation rate by the masking of recessive or nearly recessive deleterious mutations.[2] However, once a diploid species reaches its own mutation selection balance equilibrium, the mutation load actually increases beyond what it was under haploidy (Hopf et al. 1988; Haldane 1937). There must be another factor in addition to diploidy that allows complex multicellular diploids to tolerate a high mutation rate.

Sex helps cope with deleterious mutation in a variety of ways: by masking deleterious recessives (Bernstein et al. 1985c; Michod 1995), by avoiding Muller's ratchet (Muller 1932, 1964), and by removing deleterious mutations from the population (Kondrashov 1988). To these factors we may add the way in which sex maintains a higher heritability of fitness (in the sense of a higher regression of adult fitness on zygote genotype) in the face of within-organism change resulting from somatic mutation. As the mutation rate increases in sexual diploid organisms, the regression of fitness on zygote gene frequency actually increases (see figure A-2). In other words,

101

as the mutation rate increases, and along with it the amount of within-organism change, more of the variance in fitness in sexual diploids is heritable, that is, explained by the alleles carried in the zygote. How can this be? The greater mutation rate must result in greater levels of within-organism change. At equilibrium this within-organism change must be balanced by a larger covariance of fitness with zygote frequency. This is what the Price equation 5-2 says. In haploid and asexual diploid populations this is accomplished by a greater variance in zygote gene frequency (figure A-5), while in sexual populations this can be accomplished by a greater regression of organism fitness on zygote frequency.

The fitness statistics studied in appendix A apply before and after the transition. It is unclear whether these equilibrium statistics can be extended into the nonequilibrium realm of evolutionary transitions and if the results will hold up under more realistic genetic models. If so, the greater precision in the mapping of cooperative propensity onto organism fitness should allow sexuals to make the transition from cells to multicellular organisms more easily under more challenging circumstances. This result is consistent with the view that the protist ancestor of multicellular life was likely sexual (Maynard Smith and Szathmáry 1995).

In single-locus haploid models, sex has no effect on within-organism variation, because, without the possibility of heterozygosity, there can be no effects of recombination in a single-locus model.[3] With diploidy there are two dominance relations to consider, at the level of the cell and at the level of the organism (see table A-1). Diploidy may facilitate the initial increase of cooperation through masking the advantage of defection in $CD$ zygotes. Diploids may also reach larger organism sizes than haploids, although at much reduced levels of cellular cooperation. If adult size is held constant, however, these advantages of diploidy no longer pertain (J. Lie and R. E. Michod, unpublished results, although some aspects of the model are described in "Evolution of Adult Size" in chapter 6). The buffering effect of increasing the regression of fitness at the organism level on zygote genotype, whereby the level of harmony and cooperation within the organism is maintained in the face of increasing within-organism change, still pertains under diploidy, although it is affected by dominance.[4]

## STRENGTHS AND WEAKNESSES OF THE MODEL

The transition in evolution from single cells to multicellular organisms involves selection at several different levels, including the cell and the group

of cells that make up the organism. The evolution of conflict and cooperation among cells within organisms has been represented in terms of several
basic parameters and variables at a single gene locus. This simplification
permits mathematical analysis of the consequences of development but the
limitations of the approach should be kept in mind.

Selection within the organism depends upon the rate of cell division and
cell death. Within-organism variation after development is represented by
the expected number of cells of different types in the adult form—the $k_i$ and
$k_{ij}$ variables defined in tables 5-2 and 5-3. Many aspects of the analysis, especially the various equilibria and their stability (equations 5-1–5-4), could
be obtained without explicitly specifying values for these variables. For the
numerical studies reported in the graphs, specific models of fitness (linear)
and within-organism variation had to be assumed (tables 5-2 and 5-3). Cell
death is included in the model but was not studied here numerically for reasons of space. Including cell death increases the levels of within-organism
change, as more cell divisions are required to achieve a given adult size.
There may be other more realistic mutation selection models for the $k_{ij}$ variables, and the model was set up as a framework with this possibility in mind.
Although only haploid asexuality is considered in the text, both diploidy
and sex are studied in appendix A, where it is found that the reproductive
mode can have profound effects on the evolution of cooperation and the
emergence and heritability of fitness at the individual level.

I assume that organism fitness is a linear function of the frequency of cooperating cells in the adult. In modern organisms with many specialized cell
and tissue types, the dependence on cell interactions is highly nonlinear,
with fitness dropping to zero if there are too few cells of a necessary tissue
type, or too many cells of a malignant tumor. However unrealistic, the assumption of linearity has several virtues. Besides being simple and suscep-
tible to analytical treatment, linearity likely underrepresents the importance
of cell-cell interaction to the organism and for this reason represents a worst
case for the evolution of cooperation. As already pointed out, the basic
structure of the model allows incorporation of other assumptions concerning the details of fitness and within-organism change.

The complexity of interaction among different cell types is represented
by a single variable—cooperativity. This assumption is primarily for reasons of simplicity; however, it is similar to the common assumption in sociobiology of, for example, representing the interactions in a wasp colony,
with different castes and functions, by studying a single cooperative strategy. This approach has led to a deep understanding of the evolution of social behavior of organisms within social groups, and I believe a similar ap-

proach will prove useful in the study of the social behavior of cells within organisms.

Although based on a probabilistic model of cell proliferation (table 5-3), the model is deterministic, because it makes use of the expected values of the different cell types, the $k$ variables, in the determination of fitness. The consequences of relaxing this assumption are difficult to predict a priori and this is a matter worthy of careful study. Stochastic changes would likely increase the variance in distribution of cell types both within and between organisms.

Because of the hierarchical nature of selection within and between organisms, there are two levels of selection at which to consider mutational effects—the cell and the organism (recall table 5-1). Mutations that benefit both the fitness of cells and the fitness of the whole organism will sweep through the population. Mutations that detract from the fitness of both levels will be lost from the population. There is some evidence for this kind of effect (Demerec 1936). We have studied such completely deleterious mutations and found that they select for the modifiers of within-organism change studied in the context of selfish mutations in the next chapter (Michod and Roze 1998). The occurrence of selection among cells within the organism may have the benefit of lowering the mutation load in the population (Crow 1970; Whitham and Slobodchikoff 1981; Otto and Orive 1995). This effect depends upon how many cells are sampled from the adult to initiate the offspring and the opportunity for selection within and between organisms (Michod and Roze 1998). Selfish mutations that benefit the cell's replication rate but detract from organism fitness are the case studied here, since they threaten the integrity of the organism in a more aggressive manner. Evidence exists for this kind of mutation in animals—most notably malignant cancer mutants. In plants, cancer is less a problem because plant cells have a cell wall and are not highly mobile. The other class of mutations that harm the cell but benefit the organism can be addressed by a simple adjustment of the parameters in the models studied here.

There is considerable evidence that within-organism mutation and selection among cells threatens the integrity of even modern organisms. Both somatic mutation (Inov et al. 1993; Kupryjanczyk et al. 1993; Shibata et al. 1994; Akopyants et al. 1995; Blumenthal 1992; Dennis et al. 1981; Farber 1984; Hague et al. 1993; Nielsen et al. 1994; Nowell 1976; Ramel 1992; Temin 1988; Tsiotou et al. 1995; Chigira and Watanabe 1994; Coppes et al. 1993; Hoff-Olsen et al. 1995; Miyaki et al. 1994; Talbot et al. 1995) and within-organism selection (Dennis et al. 1981; Nowell 1976; Temin 1988;

Gatenby 1991; Michelson et al. 1987) are critical in the development of many human cancers. Human somatic mutation rates in tissue culture are similar to rates of naturally occurring mutations in the germ line when expressed on a per cell division basis (Kuick et al. 1992). Consequently, I expect that in nature—in the presence of the many naturally occurring sources of DNA damage and mutation—the rates of somatic mutation will be higher than germ line rates on a per cell division basis. Evidence for selection among mutant somatic cells exists in plants (Whitham and Slobodchikoff 1981; Gaul 1958; Stewart 1978; Stewart et al. 1972), where somatic mutation and selection create genetic mosaics (Whitham and Slobodchikoff 1981; Klekowski and Kazarinova-Fukshansky 1984). Selection among cells within the organism depends upon cells being able to express their own characteristics based on their own genotype, as has been observed in plants (Stewart 1978; Stewart et al. 1972). Somatic selection is likely to be strong in modular organisms that undergo continuous mitotic proliferation and/or develop by budding, like corals, aspens, creosote, or *Hydra* (Hughes 1989). Cellular selection may be an important defense against aging in constantly replicating cell lineages (Bernstein and Bernstein 1991). For example, blood-forming cells and the epithelial cells that line the intestines replicate continuously and do not appear to age, while liver, brain, or muscle cells do not divide once they are fully differentiated and do age. As a result of within-organism selection, asexual plants may be able to cope with DNA damage and live for a long time (Michod 1995).

In modern multicellular organisms there is a dual inheritance system, genetic and epigenetic (Maynard Smith 1990; Maynard Smith and Szathmáry 1995; Jablonka and Lamb 1995). During development differentiated cell types are generated by turning on and off different genes in different cells. This epigenetic state is passed on during cell reproduction so that, say, liver cells (once differentiated) give rise to liver cells. Deleterious mutation during development may then involve, not just mistakes at the loci determining cell type (as I have considered), but also errors in the epigenetic state, that is, in turning on or off the wrong genes. The situation I consider in my model involves only genetic inheritance. Because I consider a single locus determining cell type, my model cannot allow for different genes to be turned on and off in different cells. A more complicated multilocus model determining cell type with epigenetic inheritance is under development. However, I suspect that including epigenetic mutations will only strengthen the arguments for conflict mediation discussed in the next chapter.

An organism is more than a group of cells related by common descent. So far in the model, there are no uniquely organismal functions, and the levels of cooperation attained are dictated by the outcome of selection at the organism (cell-group) and cell levels. Cells have not resigned their evolutionary rights in favor of the organism. Nevertheless, the model provides a well-defined theoretical framework for exploring these issues as a means to understanding the emergence of individuality. Let us now consider the problem of individuality during the transition to multicellular life.

# Rediscovering Individuality

## EVOLUTIONARY INDIVIDUALS

The evolution of multicellular organisms is the premier example of the integration of lower evolutionary levels into a new, higher-level individual. Explaining the transition from single cells to multicellular organisms is a major challenge for evolutionary theory.

Individuality is an issue of central concern for all the sciences; each field must struggle with defining and then understanding its most basic units and levels. In the archaic sense, the term "individual" means indivisible: individuals cannot be divided into smaller parts, or to put it the other way, the whole is more than the sum of the parts. In philosophy, the term "individual" refers to entities that exist continuously in both space and time. Buss uses the term to mean a "physiologically discrete organism" (Buss 1987, p. viii). All of these uses of "individuality" apply in biology; however, in evolutionary biology the term "individual" is also used to refer to a level or unit of selection (Hull 1981). Darwin (1859) argued that natural selection requires heritable variations in fitness. Levels in the biological hierarchy—genes, chromosomes, cells, organisms, kin groups, groups—possess these properties to varying degrees, according to which they may function as evolutionary individuals or units of selection (Lewontin 1970; Hull 1981).

Ever since Darwin (1859), we have understood that a unit of selection must possess heritable variations in fitness or else it cannot evolve adaptations at its level of organization. To understand the origin of individuality, therefore, we must understand how the properties of heritability and fitness variation emerge at a new and higher level from the organization of lower-level units, these lower-level units being units of selection in their own right initially. For example, unicellular organisms enjoyed a long evolutionary history before they merged to form multicellular organisms. In so doing, single cells relinquished their evolutionary heritage in favor of the organism. Why and how did this occur?

To understand the origin of organisms it is helpful to think about them as groups of cooperating cells related by common descent as we did in the last chapter. Selection and interaction among cells—below the level of the organism—could destroy the harmony within the organism and threaten its individual integrity. Competition among cells might favor defecting cells that pursue their own interests at the expense of the organism. For the organism to emerge as an individual, or evolutionary unit, ways must have been found of regulating the selfish tendencies of cells, while at the same time promoting their cooperative interactions for the benefit of the organism. In addition, ways must have been found to ensure the heritability of the properties of these ensembles of cells so that the organism could continue to evolve as an evolutionary unit. The purpose of the present chapter is to understand how the transition from groups of cells to the multicellular individual, or organism, might have occurred.

For the organism to emerge as a new unit of selection, within-organism change and interaction must be controlled so that heritability of fitness can increase at the organism level. How might evolution modify the parameters of within-organism change so as to increase the fitness of the organism? According to Buss (1987), the individual integrity of complex animal organisms is made possible by the germ line—the sequestered cell lineage set aside early in development for the production of gametes. By sequestering a group of cells early in development, the opportunity for variation and selection is limited. As a consequence, evolution depends upon the fitness of organisms and the covariance of adult fitness with zygote genotype, and not the fitnesses of the cells that comprise the organism. The heritability of organism traits encoded in the zygote is thereby protected. The trait of interest concerns the level of specialization and differentiation among cells within organisms, which is represented here by the level of cooperativity among the cells.

Maynard Smith and Szathmáry argue that close kinship among cells should be sufficient to regulate the selfish tendencies of cells in an organism (Maynard Smith and Szathmáry 1995). By often reproducing through a single-cell stage (the zygote) organisms insure close genetic relatedness among their component cells. Another hypothesis argues that organisms evolve means of directly "policing" the selfish tendencies of their component cells, thereby reducing the benefits of defection even if this costs the organism (Boyd and Richerson 1992; Frank 1995). Examples of self-policing may include immune system responses and programmed cell death.

Missing from these discussions is a quantitative framework for evaluating and comparing the different hypotheses. The model developed in the last chapter can be extended to study these and other ideas about the evolution of individuality and the means by which organisms may reduce the potential for conflict among their cells.

## TWO-LOCUS MODIFIER MODEL

I now consider a second locus which is assumed to modify the parameters of within-organism change at the cooperate/defect $(C,D)$ locus studied in the previous chapter (see table 6-1 for the additional terms for the modifier model). All models make assumptions about the processes they study. In the case of the previous chapter, these assumptions concern parameters representing within-organism mutation and cell-cell interaction during development. The purpose of the modifier model is to study how evolution may modify these assumptions, specifically concerning how the cells interact, and if this can facilitate the transition to a higher level of selection. Unlike the classical use of modifier models, say, in the evolution of dominance and recombination, the modifiers studied in the present chapter are not neutral. Instead, they have direct effects on fitness at the cell and organism levels. By molding the ways in which the parameters interact so as to reduce conflict among cells, for example, by segregating a germ line early or by policing the selfish tendencies of cells, the modifiers construct the first true emergent organism-level functions.

The modifier locus is interpreted here as either a germ line locus or a mutual policing locus, according to whether the modifier allele $M$ affects the way in which cells are chosen for gametes (germ line modifier), or the parameters of selection and variation at the organism and cell level (policing modifier). A germ line modifier is assumed to sequester a group of cells with shorter development time and possibly a lower mutation rate than the soma. A self-policing modifier causes the organism to spend time and energy monitoring cell interactions and reducing the advantages of defection at a cost to the organism. In either case, the modifier locus is assumed to have two alleles $M$ and $m$, and no mutation (at the modifier locus). A group of cells expressing allele $m$ is assumed to have the same properties as the cell-groups studied in the previous chapter. A group of cells expressing allele $M$ is assumed to have different properties that represent the ideas of a germ line or self-policing.

The additional terms and definitions for the two-locus modifier model are

| Term | Definition |
|------|-----------|
| $i,j$ | Index for genotype 1, 2, 3, 4 = $CM, Cm, DM, Dm$ |
| $k_{ij}$ | Number of $i$ cells in the adult stage (soma) of a $j$-zygote |
| $k_j$ | Total number of cells in adult stage (soma) of $j$-zygote |
| $K_{ij}$ | Number of $i$ cells in the germ line of a $j$-zygote |
| $K_j$ | Total number of cells in the germ line of $j$-zygote |
| $W_j$ | Individual fitness of $j$-zygote: $W_j = k_j + \beta(k_{1j} + k_{2j})$ |
| $r$ | Recombination rate between $C/D$ and $M/m$ loci |
| $x_j$ | Frequency of two-locus $j$ genotype in total population |

*Table 6-1*

Additional notation and terms for the two-locus modifier model. The genotype frequencies, $x_j$, are measured at the gamete stage, before mating, meiosis, development and within and between organism selection. The table considers germ line modifiers that create a separate germ and somatic line, each of which may have different numbers of cell types (the $K_{ij}$ and $k_{ij}$ variables). In the case of self-policing modifiers, there is no distinction between the germ line and the soma (all cells are potential germ line cells).

given in table 6-1. Some explanation of the haploid life cycle considered here may be helpful. In the diploid life cycle, a generation typically begins with the diploid zygote, followed by development, within- and between-organism selection, the adult stage, and meiosis and the production of haploid gametes that fuse to produce the zygotes of the next generation (see, for example, chapter 4 or appendix A). In previous analyses of the two-locus haploid modifier model studied here, it was also assumed that the generation began with the zygote stage (Michod 1996; Michod and Roze 1997). In the haploid life cycle, the haploid zygote stage is followed by development, within- and between-organism selection, the adult stage (still haploid), production of haploid gametes, fusion of gametes to produce a transient diploid stage that undergoes meiosis to produce the haploid zygotes of the next generation. Although it does not affect the results, the two-locus recurrence equations are simpler for the haploid life cycle, and easier to understand, if we begin a generation with the gametes and let the $x_j$ variables be the frequencies of the different genotypes among gametes (instead of among zygotes). Consequently, the following sequence of events in each generation is assumed: gametes, fusion of gametes to make the transient diploid stage, meiosis and recombination to form haploid zygotes, development, within-

and between-organism selection to create the adult stage, and then formation of the gametes to start the next generation.

The recurrence equations for the two-locus haploid life cycle are constructed as follows. Recombination may change the genotype frequencies according to the level of linkage disequilibrium in the population. Linkage disequilibrium measures the statistical association between the frequencies at the two loci. If $G = 0$, the joint distribution of alleles in gametes is the product of the allele frequencies, and recombination has no effect on the genotype frequencies. In equation 6-1, I define linkage disequilibrium between the locus determining cell behavior and the modifier locus (using the variable $G$ for gametic phase imbalance).

$$G = x_1 x_4 - x_2 x_3 \qquad (6\text{-}1)$$

After meiosis, the frequencies of the four genotypes may change from what they were before fusion, depending on the rate of recombination ($r$) and the level of linkage disequilibrium, $G$. The new frequencies after recombination are $x_1 - rG$, $x_2 + rG$, $x_3 + rG$, and $x_4 - rG$, for the $CM$, $Cm$, $DM$, and $Dm$ genotypes, respectively. It is upon these new genotype frequencies that selection and mutation operate, as shown in equations 6-2. The full two-locus dynamical system is given in equation 6-2,

$$x_1' \overline{W} = (x_1 - rG)W_1 \frac{K_{11}}{K_1}$$

$$x_2' \overline{W} = (x_2 + rG)W_2 \frac{K_{22}}{K_2}$$

$$x_3' \overline{W} = (x_3 + rG)W_3 + (x_1 - rG)W_1 \frac{K_{31}}{K_1} \qquad (6\text{-}2)$$

$$x_4' \overline{W} = (x_4 - rG)W_4 + (x_2 + rG)W_2 \frac{K_{42}}{K_2}$$

with $\overline{W} = (x_1 - rG)W_1 + (x_2 + rG)W_2 + (x_3 + rG)W_3 + (x_4 - rG)W_4$ and linkage disequilibrium given by equation 6-1. Only three of the equations are independent, since the frequencies must sum to one. Equations 6-2 give the same results as the equations studied previously (equations 1 and 2 of Michod 1996 and in Michod and Roze 1997). The primary difference is that genotype frequencies are measured among gametes and not zygotes, and this allows a simpler form of the recurrence equations.

111

## Model Parameters

Recall the parameters of the model studied in the previous chapter. These parameters describe fitness effects at the organism ($\beta$) and cell level ($b$, $s_C$, $s_D$), the mutation rate per cell division ($\mu$), and the development time ($t$). The parameter $\beta$ is the benefit to organism fitness of cooperation among its component cells (this benefit is assumed to depend linearly on the frequency of cooperating cells in the organism's adult stage); $b$ is the benefit to cells of their not cooperating (this benefit is assumed to be expressed in terms of the cell's increased rate of replication and is expressed relative to the rate of replication of cooperating cells so that $b > 1$, since cooperation is assumed to be costly at the cell level); $s_C$ and $s_D$ are the survivals of $C$ and $D$ cells (assumed equal here for simplicity); $\mu$ is the rate of mutation per cell division during development (interpreted here as the genome-wide deleterious mutation rate at all loci leading to a loss of cell function); and $t$ is the time available for development (development allows for within-organism change resulting from mutation, $\mu$, and selection, $b$, $s_C$, $s_D$, at the cell level). A new parameter is needed in the case of sexual reproduction: $r$ is the rate of recombination between the cooperate/defect locus and the modifier locus. Again, the complexity of interaction among different cell types and tissue functions is assumed to be represented by two kinds of interactions— cooperate and not cooperate or defect. Heritability of fitness at the organism level is measured by the regression of offspring fitness on adult fitness.

A per cell division mutation rate of $\mu = 0.003$ is assumed in many of the studies reported in this chapter. As discussed in the last chapter, this rate is equal to the genome-wide mutation rate in modern DNA-based microbes (Drake 1991). Because this rate holds for organisms as diverse as bacteria, yeast, and a filamentous fungus, I think it is reasonable to assume that a similar rate likely held for those unicells that first formed multicellular groups. Nevertheless, small mutation rates also result in the evolution of modifiers of within-organism change (that is, the transition from equilibrium 3 to 4 discussed below; see, for example, panels (D) and (E) of figure 6-1).

## Equilibria of the System

The transition to individuality involves two general steps. First, cooperation must increase. As discussed in the last chapter, the increase of cooperation within the group is accompanied by an increase in the level of within-group change and a loss of cooperation. Second, modifier genes appear that regulate this within-group conflict. Only after the evolution of modifiers of within-group change do I refer to the group as an "individual"; then the group is indivisible because it possesses higher-level functions that protect its integrity. The different equilibria of the system are given in table 6-2, assuming linkage equilibrium ($G = 0$). Equilibria with $G > 0$ are discussed elsewhere (Michod and Roze 1998). The evolution of cooperation corresponds to equilibrium 3 and the evolution of the modifier corresponds to equilibrium 4. Consequently, the question of the transition to individuality boils down to the conditions for a transition from equilibrium 3 to equilibrium 4.

The evolution of functions to protect the integrity of the organism is not possible if there is no conflict among the cells in the first place. It is conflict itself (at equilibrium 3) that sets the stage for a transition between equilibria 3 and 4 and the evolution of individuality. Consider the initial situation in which the population does not have the $M$ modifier allele, so that the $m$ allele is fixed and the groups of cells have the properties studied in the previous chapter.

## Evolution of the Germ Line

The essential feature of a germ line is that gamete-producing cells are sequestered from somatic cells early in development. Consequently, gametes have a different developmental history from cells in the adult form (the soma) in the sense that they are derived from a cell lineage that has divided fewer times with, perhaps, a different mutation rate per cell replication. The main parameters influencing the evolution of the germ line are the reduction in development time in the germ line relative to the soma, $\delta$ ($t_M = t - \delta$), and the deleterious mutation rate in the germ line, $\mu_M$. Before proceeding, we must consider where the germ line cells come from. We might imagine that the total number of cells is conserved and allocated between the germ line and the soma. Cells sequestered in the germ line are no longer available for somatic function, and, for this reason, the germ line allele may detract from adult organism function. One way of representing this cost is by subtracting the germ line cells from the somatic cells in the adult form ($k_{ij} - K_{ij}$ in table 6-3). The number of cells is directly related to the development time. So if $\delta$ is very small, then the number of cells in the germ line is quite large, leaving few cells for somatic function. For this reason, we find that the transition from equilibrium 3 to 4 cannot occur by a continuous increase of $\delta$ from zero. There is a threshold value of $\delta$ resulting from the cost

| Eq. | Description of Loci | Interpretation |
|---|---|---|
| 1 | No cooperation; no modifier | *Single cells*, no organism |
| 2 | No cooperation; modifier fixed | Not of biological interest, never stable |
| 3 | Polymorphic for cooperation and defection; no modifier | *Group of cooperating cells*: no higher-level functions |
| 4 | Polymorphic for cooperation and defection; modifier fixed | *Individual organism*: integrated group of cooperating cells with higher-level function mediating within organism conflict |

*Table 6-2*

Equilibria for the modifier model, assuming $G = 0$. See appendix A, especially table A-3 and A-4, for a mathematical description of the equilibria and eigenvalues.

of the germ line which must be overcome for the modifier to increase. If there were no cost, the germ line allele would always spread, because it always increases the heritability of fitness.

The equilibria of equation 6-2 correspond to different evolutionary outcomes. Regions of stability of the different equilibria for different parameter values are given in figure 6-1 in terms of the reduction of development time caused by the germ line modifier, $\delta$, and the advantage of defection at the cell level, $b$. The transition involving the increase of cooperation (Eq. 1 to Eq. 3 in figure 6-1) has been considered previously (Michod 1997a). This transition occurs for parameter values in the regions marked Eq. 3 in the panels in figure 6-1, while cooperation will not increase for parameter values in the Eq. 1 region. The transition that interests us here involves the increase of modifiers of within-group change, and this transition occurs after cooperation increases (Eq. 3 to Eq. 4 in figure 6-1). The conditions under which the population evolves from equilibrium 3 to 4 were studied previously (Michod 1996). This transition occurs for parameter values in the Eq. 4 regions in figure 6-1. In a subsequent section, I study the components of this transition in terms of the emergence of fitness and the heritability of fitness at the new organism level.

The parameters have understandable effects on the regions of stability of the different equilibria described in table 6-2. For example, as the benefit of cooperation to the group increases from $\beta = 3$ to $\beta = 30$, larger replication benefits of defection at the cell level are tolerated, as shown in panels (A)

| Adult Stage | | Genotype of Zygote | | | |
|---|---|---|---|---|---|
| | | $CM$ | $Cm$ | $DM$ | $Dm$ |
| Germ Cells, $K_{ij}$ | $CM$ | $2^{ctM}\left(1-\mu_M\right)^{ctM}$ | | | |
| | $Cm$ | $0$ | | | |
| | $DM$ | $\dfrac{\mu_M 2^{bctM} - 2^{ctM}\left(1-\mu_M\right)^{ctM}\mu_M}{-1+2^{b-1}+\mu_M}$ | | | |
| | $Dm$ | $0$ | | | |
| Somatic Cells, $k_{ij}$ | $CM$ | $2^{ct}\left(1-\mu\right)^{ct}$ | $0$ | $0$ | $0$ |
| | $Cm$ | $0$ | $2^{ct}\left(1-\mu\right)^{ct}$ | $0$ | $0$ |
| | $DM$ | $\dfrac{\mu 2^{bct} - 2^{ct}\left(1-\mu\right)^{ct}\mu}{-1+2^{b-1}+\mu}$ | $0$ | $2^{bct}$ | $0$ |
| | $Dm$ | $0$ | $\dfrac{\mu 2^{bct} - 2^{ct}\left(1-\mu\right)^{ct}\mu}{-1+2^{b-1}+\mu}$ | $0$ | $2^{bct}$ |

*Table 6-3*

Numbers of different cell types in the soma and gamete stages for germ line modifiers. The zygote genotype is given across the top (column) and the genotype of the cells after development is given down the rows. For genotypes containing the $m$ allele, there is no germ line, so there is no difference between the germ line stage and the somatic adult stage. The germ line is ignored in $D$-containing zygotes since by assumption there is no mutation from $D$ to $C$ and so no within-organism variation in $D$-containing zygotes. See also figure 5-1. No cellular selection in the germ line may be assumed by setting $b = c$ in the $K_{DM.CM}$ entry in the table. These issues are explored in Michod and Roze (1998). The results presented in the book assume selection in the germ line (as does the table) because this makes matters more difficult for cooperation and the modifier.

and (B) of figure 6-1. Likewise, as the size of the organism decreases from $t = 40$ to $t = 20$ to $t = 10$ (shorter development times), larger benefits of defection at the cell level are tolerated although the reduction in development time for the germ line to evolve is about the same (see panels (A), (C), and (F) of figure 6-1). As the mutation rate decreases from $10^{-3}$ to $10^{-4}$ to $10^{-5}$, it becomes more difficult for germ line modifiers to increase, in the sense that larger reductions in development time are required for a transition from equilibrium 3 to 4 (see panels (C), (D), and (E) of figure 6-1; see also Michod 1997a, figure 1).

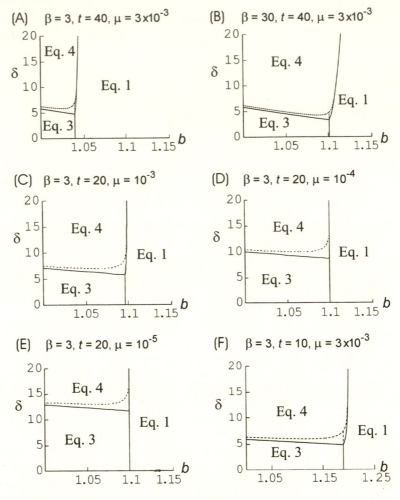

*Figure 6-1*

Stability of evolutionary equilibria for germ line modifiers. The modifier is assumed to decrease the development time for the germ line (when compared to the soma) by an amount δ. Regions of stability for the different equilibria described in table A-3 are given as a function of the advantage of defection, *b,* (abscissa), and δ,(ordinate) for different values of the mutation rate, μ, development time, *t,* and advantage of cooperation, β. Solid curves are for asexual reproduction and dashed curves for sexual reproduction, assuming a recombination rate of $r = 0.25$ between the modifier and cooperate/defect locus. Cells sequestered in the germ line are not available for somatic function. The mutation rate is assumed to be the same in the soma and the germ line. See also figure 1 of Michod (1997a) for a more detailed treatment of the boundary between equilibrium 3 and 4 in three dimensions (*b,* μ, δ). Adapted from Michod and Roze 1997.

Figure 6-1 shows the limits of these regions in $(b,\delta)$ space, for different values of the time for development $t$, the benefit of cooperation, $\beta$, and the mutation rate, $\mu$. In the case of asexual reproduction these different regions do not overlap. This is usually the case with sex, but with sex there can be regions in which more than one equilibrium are stable at the same time. For example, there can be a small region at the boundary between equilibrium 1 and 4 where these two equilibria are both stable. This suggests that with sex, mixed populations of individual cells (no organisms, Eq. 1) and organisms interpreted as well-integrated groups of cells (Eq. 4) may coexist. However, the transition from equilibrium 1 to equilibrium 4 is less interesting in terms of an evolutionary scenario towards individuality, because it supposes the simultaneous appearance of $C$ and $M$ alleles in the population. It is more reasonable to consider the evolutionary transition via equilibrium 3. Consequently, the boundary that interests us is the one between the regions of stability of equilibrium 3 and 4. With sex, a narrow region may exist between equilibrium 3 and 4, in which neither equilibrium 3 or 4 is stable, but where a new equilibrium with $G > 0$ exists and groups with and without the modifier may co-exist (Michod and Roze 1998).

Figure 6-1 concerns mutations that are deleterious at the organism level and beneficial $(b > 1)$ or neutral $(b = 0)$ at the cell level. When mutations are deleterious at the cell level $(b < 1)$, the curves in all panels of figure 6-1 continue to increase ($\delta$ gets larger) as $b$ decreases, until at $b = 0$ (mutations do not replicate but remain in the organism) the curves reach a finite value of $\delta < t$ (Michod and Roze 1998). In other words, as mutations become more deleterious at the cell level, the cost of the germ line must be smaller (in terms of the number of cells committed to the germ line) for a germ line to evolve. As mutations range from being advantageous $(b > 1)$ to neutral $(b = 1)$ to disadvantageous $(b < 1)$ at the cell level, within-organism change is less of a threat, and the evolution of a germ line becomes more difficult. Nevertheless, germ line modifiers may still evolve for the three kinds of mutations (similar results hold for the case of self-policing modifiers considered below). The advantage of a germ line, and of conflict mediation generally, depends upon the germ line increasing the heritability of fitness for the organism, and this advantage pertains (to varying degrees) to all three kinds of mutations (Michod and Roze 1998).

As already discussed, the transition to individuality via equilibrium 3 involves two steps: initial increase of cooperation within the group, and concomitantly increase of the level of within-group change, since mutation leads to loss of cooperation, and then appearance of the germ line to regulate this within-group conflict. Only after the evolution of modifiers of within-group conflict can we refer to the integrated group of cooperating cells as an "individual," since the group is now indivisible as it possesses higher-level functions that protect its integrity.

## EVOLUTION OF THE MUTATION RATE

Modifiers lowering the mutation rate ($\mu_M < \mu$) are also selected for in this model. Maynard Smith and Szathmáry suggest that germ line cells may enjoy a lower mutation rate but do not offer a reason why (1995). Bell interpreted the evolution of germ cells in the Volvocales as an outcome of specialization in metabolism and gamete production to maintain high intrinsic rates of increase while algae colonies got larger in size (Bell 1985; Bell and Koufopanou 1991; Maynard Smith and Szathmáry 1995, pp. 211–213). I think there is a connection between these two views.

As metabolic rates increase so do levels of DNA damage. Metabolism produces oxidative products that damage DNA and lead to mutation. It is well known that the highly reactive oxidative by-products of metabolism (for example, the superoxide radical $O_2^-$ and the hydroxyl radical $\cdot OH$ produced from hydrogen peroxide $H_2O_2$) damage DNA by chemically modifying the nucleotide bases, inserting physical cross-links between the two strands of a double helix, or breaking both strands of the DNA duplex altogether. The deleterious effects of DNA damage make it advantageous to protect a group of cells from the effects of metabolism, thereby lowering the mutation rate within the protected cell lineage.

This protected cell lineage—the germ line—may then specialize in passing on the organism's genes to the next generation in a relatively error-free state. Other features of life can be understood as adaptations to protect DNA from the deleterious effects of metabolism and genetic error (Michod 1995): keeping DNA in the nucleus protects the DNA from the energy-intensive interactions in the cytoplasm, nurse cells provision the egg so as to protect the DNA in the egg, sex serves effectively to repair genetic damage while masking the deleterious effects of mutation. The germ line may serve a similar function of avoiding damage and mutation: by sequestering the next generation's genes in a specialized cell lineage these genes are protected from the damaging effects of metabolism in the soma.

As just mentioned, according to Bell (1985), the differentiation between the germ and the soma in the Volvocales results from increasing colony size, with true germ soma differentiation occurring only when colonies reach about $10^3$ cells as in the *Volvox* section *Merillosphaera*. Assuming no cell death, this colony size would require a development time of approximately

118

$t = 10$ in my model (in reality, because of cell death, larger $t$ with more risks of within-colony variation would be needed to achieve the same colony size). Although Bell interpreted the dependence of the evolution of the germ line on colony size as an outcome of reproductive specialization driven by resource and energy considerations, this relation is also explained by the need for regulation of within-colony change (see panel (F) of figure 6-1). As discussed in the section "Origin of Multicellular Life" towards the end of this chapter, the ancestors of modern metazoans were composed of approximately $10^3$ cells and likely evolved "type 1" development as a means of conflict mediation (Blackstone and Ellison 1998).

## EVOLUTION OF SELF-POLICING

Another means of reducing conflict among cells is for the organism to actively police and regulate the benefits of defection (Boyd and Richerson 1992; Frank 1995). How might organisms police the selfish tendencies of cells? The immune system and programmed cell death are two possible examples. There are several introductions to the large and rapidly developing area of programmed cell death, or apoptosis (Carson and Ribeiro 1993; Ameisen 1996; Anderson 1997). To model self-policing, I let the modifier allele affect the parameters describing within- and between-organism selection and the interaction among cells. Within-organism selection is still assumed to result from differences in replication rate, not cell survival, by assuming $s_C = s_D = 1$. Cooperating cells in policing organisms spend time and energy monitoring cells and reducing the advantages of defection to $b - \varepsilon$ at a cost to the organism, $\delta$. The parameter $\delta$ is now completely different from the germ line modifier $\delta$; $\delta$ is now the fitness cost of self-policing at the organism level. To sum up, $m$ genotypes are described by the parameters $b$ and $\beta$, while in $M$ genotypes, the benefit of cooperation becomes $\beta - \delta$ and the benefit of defection becomes $b - \varepsilon$.

Figure 6-2 shows the regions of stability of different equilibria given in table 6-2 as a function of $b$ and $\delta$, for several values of development time, $t$, and benefit of cooperation at the whole-organism level, $\beta$. The modifier increases (equilibrium 4 is stable), if the cost of policing, $\delta$, does not exceed the boundary between regions Eq. 3 and Eq. 4 in figure 6-2.

In figure 6-2, there is a threshold level of the benefit of defection, $b$, above which the organism cannot be maintained, with or without the modifier; equilibrium 1 is stable in this region. This region is defined as the nearly vertical line of Eq. 1 in figure 6-2. This threshold increases, permitting greater levels of defection, as the development time decreases (compare

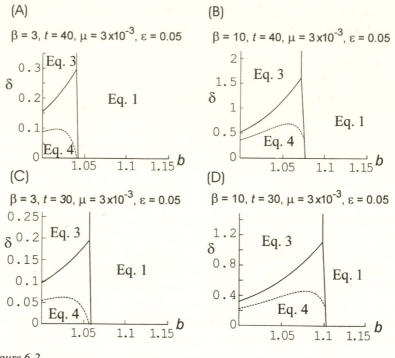

(A)

$\beta = 3$, $t = 40$, $\mu = 3 \times 10^{-3}$, $\varepsilon = 0.05$

(B)

$\beta = 10$, $t = 40$, $\mu = 3 \times 10^{-3}$, $\varepsilon = 0.05$

(C)

$\beta = 3$, $t = 30$, $\mu = 3 \times 10^{-3}$, $\varepsilon = 0.05$

(D)

$\beta = 10$, $t = 30$, $\mu = 3 \times 10^{-3}$, $\varepsilon = 0.05$

*Figure 6-2*

Stability of evolutionary equilibria for self-policing modifiers. Regions of stability of the different equilibria as a function of the advantage of defection, $b$, and the cost of policing, $\delta$, for different values of the development time, $t$, and benefit of cooperation to the organism, $\beta$. The equilibria are described in table 6-2. The modifier is assumed to decrease the advantage of defection by an amount $\varepsilon$ ($\varepsilon = 0.05$ in all panels). Solid curves are for asexual reproduction and dashed curves for sexual reproduction, assuming a recombination rate of $r = 0.25$. See figure 3 of Michod (1996) for a more detailed treatment of the boundary between equilibrium 3 and 4 in three dimensions ($b$, $t$, $\delta$). Based on results of D. Roze and R. E. Michod.

panel (C) with panel (A) and panel (D) with panel (B)). Once the modifier evolves (region Eq. 4), greater levels of defection are tolerated as the threshold slants to the right. The effect is more pronounced for higher levels of $\beta$ ($\beta > 10$; results not shown, but one can see the general effect by comparing panels (B) and (A) of figure 6-2). This effect also occurs in the case of the germ line modifier (see panel (B) of figure 6-1).

As the benefit of defection begins increasing, larger costs of policing are tolerated and the modifier still increases (boundary between Eq. 3 and Eq.

4 tends upward as $b$ increases from 1). Recombination (dashed curve) reduces the prospects for the policing modifier, as it did in the case of germ line modifiers (figure 6-1), although the effect is larger in magnitude in the case of the policing modifier. This effect of recombination becomes most pronounced as the Eq. 1 threshold is reached (figures 6-1 and 6-2), leading to the humped curve defining the boundary between Eq. 3 and Eq. 4.

There are important differences in how self-policing and germ line modifiers are modeled. In the case of the germ line, both the cost and the benefit of the $M$ allele vary with $\delta$, the reduction in development time in the germ line. The cost of the germ line increases as $\delta$ decreases. In the model of self-policing, the cost and the benefit are independent. In the graphics of figure 6-2, the benefit of defecting at the cell level is fixed at $\varepsilon = 0.05$ (the replication rate of the $D$ cells is lowered by 5%), while the cost to the organism of policing defecting cells, $\delta$, is given on the $y$-axis.

## EVOLUTION OF ADULT SIZE

Many factors influence adult size. In the present context, defecting cells are assumed to replicate faster and so produce larger, though less functional, adults. Because organism fitness is assumed to depend upon the size of the adult, in addition to the level of cooperation among its component cells, there is an advantage of defection at the organism level in addition to its advantage at the cell level (recall table 5-1). One way of reducing the temptation of defection is to control adult size, thereby removing the advantage of defection (cost of cooperation) at the organism level. Defecting cells still have a selection advantage within organisms; fixing adult size only removes the positive effect of defection on size at the organism level.

Jie Lie and I have considered an extension of the discrete-generation model studied here (equations 5-1 and 6-2), in which a constant adult size is attained for all groups by assuming that the different kinds of zygotes develop for different periods of time (Lie 1998). For example, we allow $C$ zygotes to develop for a longer period of time than $D$ zygotes, so that both have the same number of cells in the adult stage (we maintain the assumption used here that $C$ cells divide more slowly than $D$ cells, $b > c = 1$). We further assume there is a fixed time for reproduction, so that $D$ zygotes reproduce for a longer period of time, since they reach adult size more quickly. Because of the exponential nature of cell growth, only small differences in development time are needed to attain a fixed adult size. Consequently, there is little difference between organisms in the time available for reproduction. We have not yet considered a model with overlapping generations, although this is clearly in need of study. Our results to date indicate that constant adult size makes it much easier for cooperation to increase and this effect is more pronounced for smaller mutation rates. Likewise, much greater levels of harmony and cooperation are maintained within the organism if adult size is regulated. We have yet to consider the effects of cell death, which should only increase within-organism conflict. Cell death may have important effects with regard to organism size. Cell death increases the number of cell divisions required to reach a given adult size, and this has the additional consequence of increasing the opportunity for within-organism change and variation.

## EFFECT OF TRANSITION ON THE LEVEL OF COOPERATION

I now consider the consequences of an evolutionary transition from equilibrium 3 to 4, in this section on the level of cooperation and synergism attained, and in the next two sections on the heritability of fitness at the new organism level. For reasons of space, I consider only the evolution of the germ line, but qualitatively similar results have been obtained for the other forms of conflict mediation, such as self-policing and determinate size.

In figure 6-3 we see that the transition dramatically increases the level of cooperation in the population. The level of cooperation always increases during a transition (if equilibrium 4 is stable). Linkage disequilibrium increases during the transition. To understand the effect of this evolutionary transition on the regulation of the within-organism change and the heritability of fitness at the new level, we need to adapt the covariance methods of Price to the present system of equations.

## INCREASE OF FITNESS COVARIANCE AT ORGANISM LEVEL

Recall the Price covariance equation 5-2 studied in the last chapter and how it partitions change to the two levels of selection. The first term of the Price equation is the covariance between fitness and genotype and affects the heritable aspects of fitness; the second term is the average of the within-organism change. The first term can be considered as representing the selection between organisms within the population, and the second term the selection between cells within the organism. When the population is at an

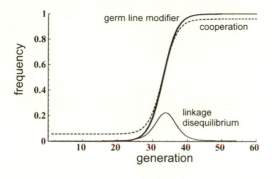

*Figure 6-3*

Frequencies of cooperation, modifier, and linkage disequilibrium during evolutionary transition. The germ line modifier refers to the $M$ allele and cooperation to the $C$ allele; asexual reproduction is assumed. The values of the parameters are $c = 1$, $t = 40$, $\delta = 35$, $b = 1.1$, $\beta = 30$, $\mu = \mu_M = 0.003$, and $r = 0$.

equilibrium, $\Delta q = 0$, and so it must be the case that the two terms on the right-hand side of equation 5-2 equal one another in magnitude $(\mathrm{Cov}_q[W_i, q_i])/\overline{W} = -\mathrm{E}_{Wq}[\Delta q_i]$.

In figure 6-4, the two components of the Price covariance equation 5-2 are plotted during the transition from equilibrium 3 to equilibrium 4 (increase in frequency of the germ line allele, from 0 to 1)(Price 1970). These components partition the total change in gene frequency into heritable fitness effects at the organism level (solid line) and within-organism change (dashed line). In the model studied here, within-organism change is always negative, since defecting cells replicate faster than cooperating cells and there is no back mutation from defection to cooperation. At equilibrium, before and after the transition, the two components of the Price equation must equal one another. This can be seen in figure 6-4 by the fact that the two curves begin and end at the same point. However, during the transition we see that the covariance of individual fitness at the emerging organism level with zygote genotype (solid curve of figure 6-4) is greater than the average change at the cell level (dashed curve of figure 6-4).

This greater covariance in fitness at the higher level forces the modifier into the population. In figure 6-4, we see that modifiers of within-organism change evolve by making the covariance between fitness at the organism level and zygote genotype more important than the average within-organism change. This suggests that the modifiers increase the heritability of fitness at the new organism level.

123

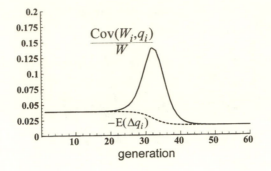

*Figure 6-4*

Study of evolutionary transition by Price equation. The parameter values are the same as in figure 6-3. Adapted from Michod and Roze 1998.

## HERITABILITY OF FITNESS AND THE EVOLUTION OF INDIVIDUALITY

Natural selection requires heritable variations in fitness. Levels in the biological hierarchy possess heritability of fitness to varying degrees, according to which they may function as units of selection. From E. O. Wilson (1975) and the study of the transition from solitary animals to societies, to Buss (1987) with the study of the transition from unicellular to multicellular organisms, and more recently Maynard Smith and Szathmáry (1995), attention has focused on understanding transitions between these different levels of selection or different kinds of evolutionary individuals. However, little attention has been given to the fundamental defining characteristic of a unit of selection, heritability of fitness. How can heritability of fitness emerge at the new level during an evolutionary transition?

Before the evolution of cooperation, in the present model, the population is composed of $Dm$ cell types (equilibrium 1 in table 6-2). In such a population the heritability of fitness equals unity, because there is either no sex, or no effect of recombination if there is sex (in $Dm \times Dm$ matings), and there is no within-organism variation or change (I assume no mutation from $D$ to $C$). When the $C$ allele appears in the population, evolution (directed primarily by kin selection) may increase its frequency, leading to greater levels of cooperation (from equilibrium 1 to equilibrium 3). With the evolution of greater cooperation, within-organism change increases, because of mutation from $C$ to $D$ and selection at the cell level. As a consequence of the evolution of cooperation and increasing within-organism change, the heritability of fitness must decrease at the organism level.

The organism cannot evolve new adaptations, such as enhanced cooperation, if these adaptations are costly to cells, without increasing the opportunity for conflict within and thereby decreasing the heritability of fitness. Deleterious mutation is always a threat to new adaptations, and it can produce cells that go their own way. By regulating within-organism change, there is less penalty for cells to help the organism. Without a means to regulate within-organism change, the "organism" is merely a group of cooperating cells related by common descent. Such groups are not individuals, because they have no higher-level functions that exist at the new organism or group level.

The existence of a zygote stage in the life cycle serves to decrease the within-organism change by increasing the relatedness among cells (Michod and Roze 1998). However, as shown in the last chapter, within-organism change can be significant even in this case. The main criterion of significance is whether within-organism variation leads to selection of modifiers to reduce it. We have found that such modifiers increase in frequency, leading to an evolutionary transition that we have interpreted as the emergence of individuality, because these modifiers represent the first higher-level functions. However, does heritability of fitness—the defining characteristic of an evolutionary individual—actually increase during the transition between equilibrium 3 and 4?

Heritability of fitness may be defined as the regression of the fitness of offspring on fitness of the parents as in table 5-2 (see, for example, Falconer 1989). When the population is at equilibrium 3 or equilibrium 4, this definition gives a simple expression for heritability in the two-locus model, as I show in figure 6-5 for the case of equilibrium 3.

An approach similar to that given in figure 6-5 shows that the heritability of fitness equilibrium 4 is equal to $K_{11}/K_1$. We always have $K_{11}/K_1 > k_{22}/k_2$, so the evolutionary transition always leads to an increase in the heritability of fitness. If we go to the eigenvalues of the different equilibria given in table A-4, we note that these eigenvalues are ratios between products of fitnesses and heritabilities. This illustrates clearly that whether a new characteristic can increase in frequency in the population is determined by the heritability of fitness of individuals with the new feature not simply individual fitness.

During the transition between equilibrium 3 and equilibrium 4, all four genotypes are present in the population. It is not possible to simplify the expression for the heritability as above (as, for example, in figure 6-5); how-

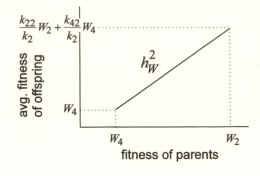

$$\frac{k_{22}}{k_2}W_2 + \frac{k_{42}}{k_2}W_4 \cdots$$

avg. fitness of offspring

$h_W^2$

$W_4$

fitness of parents

$W_4 \qquad W_2$

*Figure 6-5*

Heritability of fitness at equilibrium 3. At equilibrium 3 the only genotypes in the population are genotypes 2 (*Cm*) and 4 (*Dm*). The regression is equal to $h_W^2 = k_{22}/k_2$. Based on unpublished results of D. Roze and R. E. Michod.

ever, the expression $h_W^2 = \text{Cov}(W_p, W_O)/\text{Var}(W_p)$ in table 5-2 can still be calculated and used, where $W_p$ is the fitness of each parent and $W_O$ is the average fitness of the offspring produced by the parents.

Initially, before the evolution of cooperation between cells, the heritability of fitness is unity. After cooperation evolves (because of high kinship), heritability is significant at the group (organism) level ($h_W^2 \approx 0.6$, figure 6-6), but this value is still low for asexual haploidy (heritability at the organism level should equal unity in the case of asexual organisms when there is no environmental variance). Low heritability of fitness at the new level resulting from significant within-organism change poses a threat to continued evolution of the organism. In the case considered in figure 6-6, development time, and hence organism size, could not increase without the evolution of conflict modifiers. Indeed, as already noted, the continued existence of cell-groups at all is highly unlikely, since the cooperation allele is at such a low frequency and stochastic events would likely lead to its extinction. Before the evolution of modifiers restricting within-organism change, the "organism" is just a group of cooperating cells related by common descent from the zygote. As the modifier begins increasing, the level of within-organism change drops (dashed curve in figure 6-6) and the level of cooperation among cells increases dramatically (dashed curve of figure 6-3), as does the heritability of organism fitness (solid curve of figure 6-6).

The essential conclusion is that even in the presence of high kinship among cells, there remains significant within-organism change. By "significant," I mean this change leads to the evolution of a means to regulate it, such as the segregation of a germ line during the development or the evo-

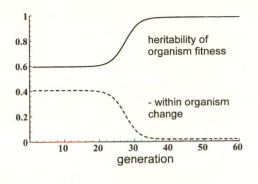

*Figure 6-6*

Heritability of organism fitness and within-organism change during evolutionary transition. The magnitude of within-organism change of cooperation is plotted; because cooperation decreases within the organism, within-organism change is negative. The parameter values are the same as in figure 6-3. Adapted from Michod and Roze 1997.

lution of self-policing. Once within-organism change is controlled, high
heritability of fitness at the new organism level is protected. Individuality
at the organism level depends on the emergence of functions allowing for
the regulation of within-organism change. Once this regulation is acquired,
the organism can continue to evolve new adaptations without increasing the
conflict among cells, as happened when cooperation initially evolved (transition from equilibrium 1 to equilibrium 3).

## SEX AND INDIVIDUALITY

Sex and individuality are in constant tension, because sex involves fusion
and mixis of genetic elements, and so naturally threatens the integrity of
evolutionary units. Yet sex is fundamental to the continued well-being of
evolutionary units too. Sex and its antithesis in the evolution of reproductive systems, parthenogenesis, provide different options for the reproduction of evolutionary units. Although sex seems to undermine individuality,
sex has been rediscovered as each new level of individuality emerges in the
evolutionary process. Sex holds the promise of a better future and a more
whole and undamaged individual. Theories for the evolution of sex are discussed in three collections of papers on the topic (Michod and Levin 1988;
Stearns 1987; American Genetics Society Symposium for the Evolution of
Sex 1993). Genetic redundancy and repair occur during the sexual cycle and
are the key to greater wholeness and well-being for the individual (Michod

1995). Cloning, on the other hand, offers ease and efficiency of reproduction at the expense of future generations and the well-being of the individual.

But with the successful cloning of a sheep, has biological science found a more direct means to perpetuate what makes us the individuals we are (Michod 1997c)? The possibility, however faint, that a person might create offspring without the benefit of a partner has brought that question and others about sexual reproduction into unusual prominence. After all, sex extracts high costs in energy, in time, and in resources. Would it not be more efficient to make copies of ourselves asexually, as some think? Does generating one new person by combining the genes of two aging parents make any more sense than a one-for-one exchange? Would begetting a clone bring about a closer approximation of immortality than procreating in the usual fashion?

For those who have fantasized—and the fantasy seems all too common— that cloning could lead to the endless renewal of individual lives, the biological evidence suggests otherwise. In fact, it turns out that sex leads to a kind of immortality by repairing the genes of the egg and sperm cells so essential for the continuation of life (Michod 1995). And cloning, on the contrary, far from being rejuvenating, could threaten the continuing evolutional well-being of genes, cells, and organisms and even the very nature of species as discussed in the next chapter in the section "Darwin's Dilemmas."

Sex also affects the conditions for an evolutionary transition to higher levels of individuality and greater complexity. In the case of germ line modifiers, sex requires larger decreases in development time in the germ line (compare the dashed curves in all panels of figures 6-1–6-6 with the solid curve, or see Michod 1996, figure 2, for the effect of recombination on evolution of the modifier). With sex it also takes longer for the transition to occur (results not shown here for reasons of space). The modifier increases by virtue of being more often associated with the more fit $C$ alleles in gametes and recombination breaks apart this association.

These effects of sex can be understood by realizing that modifier allele ($M$) increases in frequency by virtue of its association with the more fit (at equilibrium 3) $C$ allele. The modifier has no effect in $D$ zygotes, and the frequency of $C$ cells in the gametes of a $CM$ organism is greater than in the gametes of a $Cm$ organism, because of the effect of the modifier. For the values of $b$ where the $C$ organisms have more offspring than the $D$ organisms (values of $b$ where equilibrium 3 is stable), the $M$ allele is favored if its cost is not too high. Recombination due to sexual reproduction has two effects: it can create more *CM* associations by recombination between *Cm* and *DM* parents, or it can break apart *CM* associations by recombination between *CM* and *Dm*. When the modifier increases, matings between *CM* and *Dm* gametes are more frequent than matings between *Cm* and *DM* gametes, and so recombination more often breaks *CM* associations than creates them. In this sense sexual reproduction makes the increase of the frequency of the *M* allele in the population more difficult, and this effect increases as the Eq. 1 boundary, at which the *C* allele can no longer be maintained, is reached.

Although sex can retard the transitions modeled here, because of the effects of recombination in breaking up the genetic associations needed for the modifier to increase, I do not see these quantitative differences as presenting any real barriers to the evolution of conflict modification and evolutionary transitions in sexual progenitors. More important, I think, is the way in which sex organizes variability and heritability of the traits and capacities that affect the fitness of the new emerging unit. These issues are studied in appendix A and the results have been briefly discussed in the previous chapter in "Effects of Sex and Diploidy on the Emerging Organism."

## ORIGIN OF MULTICELLULAR LIFE

My goal in developing a model of individuality has been primarily heuristic, to help us understand and think about the transition from single cells to multicellular organisms. Recently, this model has been applied (Blackstone and Ellison 1998) to the origin of multicellular animals in the late pre-Cambrian some 600 million years ago (Davidson et al. 1995; Ransick et al. 1996). Fossil evidence of the developmental forms of such early ancestors to metazoans has recently been reported (Xiao et al. 1998).

The ancestors of modern metazoans were likely microscopic forms similar to modern marine larvae that have "type 1" development, which is a widespread and basic mode of embryogenesis in modern animals. The entire ontogeny of the ancestors of multicellular animals was probably similar to the type 1 embryogenesis occurring in modern groups (Davidson et al. 1995). In type 1 development, cleavage begins immediately after fertilization and proceeds for a species-specific set number of divisions, often in the range $10 \pm 2$, as occurs during the development of sea urchin embryos. By the end of cleavage, all the blastomeres have been specified. The resulting embryo consists of approximately one thousand cells and is divided into a set of polyclonal lineages in which each element gives rise to a certain differentiated cell type or types. The cell lineage of each morphological structure in the complete embryo is typically invariant within species and follows from the positions of the successive cleavage planes with reference to each other. Determination of cell type occurs before the embryonic cells acquire any capability for cell motility.

Type 1 development is readily interpretable, using the parameters of the model of individuality developed here, as a means of reducing the time available for development, $t$, and the replication advantage of defecting cells, $b$, by maternal control of cell type (Blackstone and Ellison 1998). Maternal control of cell behavior was proposed by Buss (1987) as a way of reducing conflict among cells, thereby making it easier for cooperation to arise. A similar process occurs in models of kin selection, in which it is easier for altruism to evolve under parental control of offspring behavior than it is under offspring control (Michod 1982). In terms of the model above, maternal control of cell behavior may be represented by assuming that the conflict modifier delays any advantage for defection, $b$, for a number of cell divisions (during which the maternal factors are diluted out). By limiting development to a small number of cell divisions under control of the maternal cytoplasm, along with strictly determined cell fate prior to the movement of any cells, the opportunity for conflict would be reduced. The developmental features that Davidson et al. (1995) ascribe to the ancestors of multicellular life likely evolved to reduce conflict among cells and promote harmony within the group (Blackstone and Ellison 1998).

Type 1 development seems to work only for building the morphological structures of small larvae, consisting of at most a few thousand cells, probably because maternal control of cell type can proceed only so far during cell replication. To build more complex macroscopic metazoans a different approach was needed. According to Davidson et al. (1995), the key breakthrough was the development of a population of undifferentiated "set-aside" cells, containing the precursors to the later cell types needed to build the structures, such as limbs, of the macroscopic organism. This would entail great risks for the organism, however, in terms of the opportunity for conflict, because any defecting mutant in the population of undifferentiated set-aside cells could become systemic and destroy the organism (Blackstone and Ellison 1998). The early approach to conflict regulation of reducing $t$ (by a small and fixed number of cell divisions) and $b$ (by maternal control of cell fate prior to cell movement) would be completely circumvented by a population of undifferentiated set-aside cells with unlimited replication potential and mobility. In primitive organisms with type 1 development, there is no need for a germ line since conflict mediation is implicit in the type 1 mode of development. However, with the evolution of set-aside cells, the evolution of germ line and other modes of self-policing modifiers becomes necessary to mediate the conflict inherent in this large population of cells of unlimited replicative ability.

This expectation is borne out since it appears that the germ line evolved in concert with the breakthrough developmental patterns that permitted large macroscopic organisms (Blackstone and Ellison 1998; Ransick et al. 1996). In modern organisms, such as the sea urchin, with maximally indirect development, the cell types in the adult body are determined after embryogenesis and development of the larva. These organisms continue to use the ancestral type 1 mode of development for building the larva and a second developmental process for building the adult form. In organisms with maximally indirect development, the germ line is not primordial but is determined after embryogenesis. This makes sense in terms of the model, because only after type 1 development is complete is there an opportunity for conflict and need for germ line sequestration. In contrast, direct development (existing in arthropods, nematodes, and chordates, in which the cell types of the adult form are specified during early cleavage of the embryo) is a derived state in which the rigid framework set down by type 1 embryogenesis has been circumvented. Primordial specification of the germ line during gastrulation occurs in organisms with direct development because defecting mutants may arise early in the developmental process and threaten the integrity of the organism.

## TRANSITIONS IN INDIVIDUALITY

The models studied here support the view that the germ line and self-policing evolved for the purpose of reducing within-organism change, and that this served to facilitate the transition between cells and organisms. The germ line functions to reduce the opportunity for conflict among cells and promote their mutual cooperation both by limiting the opportunity for cell replication (Buss 1987) and by lowering the mutation rate (Maynard Smith and Szathmáry 1995). Mutual policing (Boyd and Richerson 1992; Frank 1995) is also expected to evolve as a means of maintaining the integrity of the organisms once they reach a critical size. Any factors that directly reduce the within-organism mutation rate are also favored.

The evolution of modifiers of within-organism change leads to increased levels of cooperation within the organism and increased heritability of fitness at the organism level. These conflict mediators are the first new functions at the organism level. An organism is more than a group of cells related by common descent; to exist, organisms require adaptations that regulate conflict within. Otherwise, continued improvement of the organism is frustrated by the creation of within-organism variation and conflict. The evolution of modifiers of within-organism change is a necessary prerequisite to the emergence of individuality and the continued well-being of the organism.

In summary, what happens during an evolutionary transition to a new higher-level unit of individuality, in this case the multicellular organism? While taking fitness away from lower-level units, cooperation increases the fitness of the new higher-level unit (cell to organism). In this way, cooperation may create new higher levels of fitness. However, the evolution of cooperation sets the stage for conflict, represented here by the increase of defecting mutants within the emerging organism. Modifiers restricting within-organism change evolve as the first higher-level functions at the organism level. Before the evolution of a means to reduce conflict among cells, the evolution of new adaptations (such as the underlying traits leading to increased cooperation among cells) is frustrated by mutants. Individuality requires more than just cooperation among a group of genetically related cells; it also depends upon the emergence of higher-level functions that restrict the opportunity for conflict and change within and ensure the continued cooperation of the lower-level units. Conflict leads—through the evolution of adaptations that reduce it—to greater individuality and harmony for the organism.

# Fitness Explanations

## OVERVIEW OF FITNESS AND NATURAL SELECTION

In this chapter I investigate the role of fitness in evolutionary explanations. First, I reconsider the role of fitness in my studies of evolutionary transitions given in the previous chapters. Then I consider the role of fitness in touchstone cases of natural selection, such as the "tautology problem," the evolution of selfing and sex, the immortality of the germ line in contrast to the mortality of the soma, the problem of altruism, heterozygote superiority, sickle cell anemia, and the existence of species as distinct groupings of individuals.

For the purpose of studying the emergence of new levels of complexity, I have studied the dynamics of natural selection in a multilevel setting. We have found that there is more to the emergence of fitness at new evolutionary levels than a simple engineering or optimization analysis of individuals might suggest. There is more to the dynamics of natural selection than increase of better-designed phenotypes. Frequency-dependent fitness effects frustrate the creation of new evolutionary units and prevent the application of simple maximization arguments, especially during evolutionary transitions.

Only units of natural selection have fitness. But units of selection, such as genes, cells, or organisms, do not exist in complete isolation, nor are they completely interdependent. Instead, they are embedded in a hierarchy of nested but partially decoupled levels, and any focal level provides both the context for lower-level units and the components for higher-level ones. Because evolutionary units (genes included) play the roles of both context and component at the same time, the dynamics of design at any level involves an interplay between the dynamics at all levels. There is some degree of isolation of the dynamics at different levels, due in large part to the evolution of conflict modifiers, and certain features of evolutionary individuals can be understood in this light (this was the topic of chapter 6). Conflict modification helps define evolutionary units, so that evolution at one level can proceed without being constantly frustrated by the increase of defecting mutants. However, evolutionary units are never completely protected from the selfish interests of their parts, nor are they completely isolated from the broader context of the groups in which they exist.

At the end of chapter 1, I adopted an adequacy criterion for understanding the role of fitness in evolutionary explanation. We are now in a position to assess this criterion: how and why did fitness originate, both in the transition from the nonliving to the living and in the emergence of new units of selection, and what is the role of fitness in evolutionary explanation, specifically in the many mathematical models of natural selection? In previous chapters, we considered the evolutionary passage from genes to gene networks, from gene networks to cells, and from cells to multicellular individuals. We used deterministic dynamical models of evolutionary change embodying Darwin's principles of natural selection. What can be said about the nature of explanation in this work, and, more specifically, what is the role of fitness in evolutionary transitions?

Fitness is a property of far-from-equilibrium systems (in a thermodynamic sense) capable of self-replication. Self-replication drives the cycle of life, in which ends produce beginnings: parents make offspring who become the parents of the next generation. The life cycle can be as simple as a bacterium going through cell division or as complex as the metazoan stages of fertilization, development, and mating (of course, these are just a few of the many stages in the metazoan life cycle).

Fitness relations connect the stages of the life cycle. The deconstruction of fitness, introduced in chapter 1 and used throughout population biology (this book included), breaks down the overall life cycle into manageable components, each with their assigned heritable capacities. There is no useful notion of fitness as an overall property of the organism. The whole point of evolutionary analysis is in the other direction, to break down the vague and undefined into explicit components and capacities, which when taken together with the environment and genetic system determine the evolutionary success of a class of individuals defined by shared traits and genes. Ordered patterns and transformations in physics and chemistry typically do not have a life cycle (endings don't create beginnings) and cannot possess fitness.[1] Fitness originated with self-replication, and ultimately its meaning and role in evolutionary explanation must be tied to the cycle of life and the replication of hierarchically nested evolutionary units through time. This fact alone compels us to take a dynamical point of view of fitness and its companion, design.

Evolutionary units appear to be designed, because of the repetitiveness of the life cycle, the predictability of the contingencies of existence, and the partial decoupling (through conflict mediation and other means) of the different levels of a selection hierarchy. Phenotypes appear to be well designed in the present, because the underlying structures and behaviors (heritable capacities) contributed to fitness relations in the past, and these relations were heritable, that is, passed from parent to offspring. However, there is really no design in biology, because there is not a plan. And how can there be design, without a plan? It is common to speak of the design of an organism (or some other unit of selection) as a shorthand way of describing its function at a particular activity. I have used this way of speaking many times in this book. The organism's properties may mimic the predictions of an engineer, but this does not mean that the organism was designed. By using the word "design" in evolutionary biology, we don't mean to suggest there is or was a plan, because there isn't one. Nor can we know whether there is an absolute measure of fit between the individual's capacities and its environment—only that the individual's structures and behaviors are adequate to maintain its life cycle for the time being. The apparent perfection of design we perceive in the many adaptations of the living world is a human projection, not a property of the world. Because there is no plan, in a fundamental sense there can be no design in biology, at least not if "design" has its common meaning as based on a plan. At the beginning of the next chapter, I consider this matter further.

## TRADING FITNESS THROUGH COOPERATION

The benefits of cooperation provide the imperative for forming new, more inclusive evolutionary units. Increments in fitness are traded among levels of selection through the evolution of individually costly yet group-beneficial behaviors. Defection, the antithesis of cooperation, is the bane of cooperative groups everywhere. Cooperative interactions create new, more inclusive evolutionary units by trading increased fitness at higher levels (the benefits of cooperation) for decreased fitness at lower levels of selection (the costs of cooperation; see, for example, tables 3-1 and 5-1). The models developed in previous chapters simultaneously considered several hierarchically nested fitness levels to understand how and why new units of selection emerge in the evolutionary process. Because cooperation reduces fitness of lower-level units while increasing fitness of higher-level units, it drives the transition to a new higher-level unit.

Although fueling the passage to higher levels, cooperation provides the opportunity for its own undoing through the frequency-dependent advantage of defection. Selfish interactions (defection) reap the benefits of coop-

135

eration while avoiding the costs and, for this reason, can be expected to spread within the cooperating group. Selfish individuals typically stand to gain more than their selfishness costs any other individual, especially when rare. The "tragedy of the commons" (Hardin 1968) leads to conflict among lower-level units, which may sabotage the viability of cooperation and the creation of new, higher levels of selection. In the final analysis, if cooperation submits to defection, nothing is gained as a result of defection's short-sighted success, for individual fitness eventually declines—another unsatisfying consequence of frequency-dependent selection. Certain conditions are required to overcome the inherent limits posed by frequency-dependent selection to the emergence of new levels of selection: kinship, population structure, and conflict mediation.

## KINSHIP AND POPULATION STRUCTURE

Kinship and population structure help to reduce the viability of selfishness and so promote evolutionary transitions. High kinship more closely aligns the interests of the lower-level units (cells in our case) so that individually costly but group-beneficial behaviors may evolve. Kinship does this by expanding the relevant fitness relation that controls spread of a behavior beyond only considerations of individual fitness. As reviewed in chapter 4, inclusive fitness effects determine the outcome of selection in populations in which kin are present. The average inclusive fitness effect defined in equation 4-19 provides the adaptive topography upon which evolution in kin-structured populations proceeds. Inclusive fitness includes the fitness effects of the interaction on the actor and, in addition, on other collateral kin, these later effects being diluted by the genetic relatedness of the recipients to the actor (recall equation 4-20). For a behavior to spread, its inclusive fitness effect must be positive. In the case of selfish behaviors, the positive effects on oneself are offset by negative effects on collateral kin.

Kinship and genetic population structure were critical components of the evolutionary transitions considered in the previous chapters. For example, self-structuring (Boerlijst and Hogeweg 1993) or passive structure (see equation 3-12) helped to explain the origin of proteins. Kinship was also a component of the passage to multicellular life. By funneling the life cycle through a single-celled stage (the zygote), the cells in a multicellular organism are close kin. Nevertheless, as we found in chapter 5, high kinship among cells is not enough for the multicellular group to become an evolutionary individual. Mutation during development and within-organism selection may open the door for conflict, as noncooperating cells gain a foothold and increase in frequency, thereby threatening the integrity of the organism. By further restricting the opportunity for selection at the lower level, conflict mediators institutionalize and enforce cooperation among lower-level units.

## CONFLICT MEDIATION

Mediating conflict by reducing lower-level change and the opportunity for selfishness is another critical component of the evolutionary transitions studied here. For example, population structure (culminating in the creation of the cell) restricts the opportunity for selfish interlopers on the hypercycle (chapter 3). By restricting the scope of selection at the cell level, germ line or self-policing modifiers enhance the fitness and integrity at the level of the cell-group, or organism (chapter 6). Conflict mediation increases the heritability of fitness at the new level while reducing opportunities for selfishness among lower levels. As a result of conflict mediation, cells end up relinquishing their evolutionary rights in favor of the group—which thereby becomes a new evolutionary individual. In this way conflict leads to greater harmony for the group.

Recall from chapters 5 and 6 that the organism could not evolve new adaptations, if these adaptations were costly to cells (such as enhanced cooperation and greater levels of synergism), without concomitantly increasing the opportunity for conflict within, thereby decreasing the heritability of fitness. Deleterious mutation is a threat to new adaptations as it produces cells that go their own way. By regulating within-organism change, there is less of a penalty for cells that help the group or the emerging organism. Without a means to mediate conflict by regulating within-organism change, the "organism" is merely a group of cooperating cells related by common descent. Such groups are not individuals, because they have no unique functions that exist at the organism level.

## RECONSIDERING FITNESS

In summary, we have found that fitness is both a cause and effect in evolutionary transitions. Increments in fitness are traded among levels of selection through the evolution of individually costly yet group-beneficial behaviors. This trade, if sustained through group and kin selection and conflict mediation, results in an increase in the heritability of fitness and individuality at the new higher level. In this way, new higher levels of fitness emerge in the evolutionary process.

There is little to support the view of the individual organism as a fitness-maximizing unit in this theory. Indeed, I think this image has long ago stopped being a useful metaphor for understanding natural selection. Most of the time, especially when selection is frequency dependent, as it always is when there are interactions and transitions in evolutionary units, it is hard to pinpoint what, if anything, is being maximized or increased. In contrast, many examples of wide generality have been given (such as the prisoner's dilemma game discussed in chapter 4) in which individual fitness is decreased by frequency-dependent natural selection. Conditions such as population structure, kinship, and learning can reverse the direction away from selfishness in a multilevel setting.

Multilevel selection is becoming more commonplace in evolutionary studies of natural selection (see the *American Naturalist* Supplement, volume 150, for a recent overview of this area). Multilevel selection at three levels—the molecule, the mitochondria, and the cell—is needed to explain the evolution and distribution of plant mitochondrial genomes (Albert et al. 1996). Multilevel selection is gaining prominence as an evolutionary framework, especially as interests shift from the properties of individual units to the processes by which these units are created. Lower-level units such as infectious agents modify the reproduction, behavior, and structure of higher evolutionary units. Consider only a few: the sex ratio distorter in *Drosophila,* *t*-allele in mice, son-killer in *Nasonia,* and the effects of the bacterium *Wolbachia* on the reproduction of arthropods (Werren 1997; Werren et al. 1988; Enserink 1997). The bacterium *Wolbachia* is likely present in 16% of all insect species.

No trait is immune from the effects of conflicts within. Reproduction is the most basic property of a living entity, yet the evolution of reproductive systems of whole organisms involves a compromise between different levels of selection. In addition to the examples just given, the evolution of gynodioecy (plant species with hermaphroditic and all female individuals) requires resolving conflicts at the nuclear and cytoplasmic levels, because nuclear genes are passed on through both pollen and ovules, but cytoplasmic genes are only passed on through ovules (Gouyon and Couvet 1985; Gouyon et al. 1991; Belhassen et al. 1991; Maurice et al. 1993, 1994; Manicacci et al. 1996). According to this view, genes for male sterility (all female plants) are predicted to arise in the cytoplasm while genes restoring male function should arise in the nucleus. The advantages to either depend upon the frequencies of each in the population. The resolution of this conflict likely has effects at the whole-organism level too, in terms of the outcrossing rate and avoidance of inbreeding depression. A multilevel selection context provides the setting for understanding the outcome of the various processes occurring on different levels.

Defining individual fitness as the expected number of gametes produced (for example, pollen plus ovules), as we have done repeatedly in this book, implicitly assumes that there is isogamy or that we are interested in nuclear genes (if we are considering a species with anisogamous sex in which no cytoplasm is passed on through the male gamete). If we were interested in traits caused by cytoplasmic genes in an anisogamous species, we would approach fitness differently and count only female gametes (ovules). In the case of gynodioecy, the reproductive phenotype of an organism (female or hermaphrodite) is caused by both nuclear and cytoplasmic genes. For this reason, in situations involving nuclear cytoplasmic conflict, fitness of the individual organism is not well defined unless the context of genetic information is specified (see appendix B, quote 15).

Organisms evolve the means to manipulate the structures and behaviors of one other, as recognized in Dawkins' concept of an "extended phenotype" (Dawkins 1982). Because of the ubiquity of manipulation, Dawkins concludes (correctly) that the organism cannot be a fitness-maximizing agent. From the fact that the organism is not a maximizing agent, Dawkins concludes, incorrectly (Michod 1983a), that only a gene can be a unit of selection. Units of selection do not have to be maximizing agents as Dawkins believes; rather, they need only satisfy Darwin's propositions of variation and heritability of fitness. Dawkins' argument for the gene as the sole unit of selection is the kind of mistake one can make by defining natural selection in terms of a single outcome, in his case producing a "maximizing agent" or "optimon" (see especially Dawkins 1982, chap. 10). As we know, there are many outcomes of natural selection; we especially should not be confused when fitness is not maximized, because of the ubiquity of frequency-dependent selection and selection at different levels in a hierarchy.

New evolutionary units of selection emerge out of the dynamics of natural selection as increments in fitness are traded between levels, these transactions eventually being enforced and institutionalized by the evolution of modifiers of lower-level change. In this theory, fitness is a dynamical concept, both as a cause and effect of evolutionary change and by virtue of its own malleability as it is traded between levels. Fitness binds, not only the life cycles of replicating units at one level, but also transitions between levels, as it assembles multiply nested units into higher levels of organization.

This is accomplished during the evolution of cooperation, as fitness is exported from lower to higher levels of selection.

Having considered evolutionary transitions in fitness in some detail, let us now consider touchstone problems in the theory of natural selection in the light of our dynamical view of fitness.

## THE "TAUTOLOGY PROBLEM"

Darwin did not use the phrase "survival of the fittest" in the first edition of the *Origin of Species* published in 1859. However, in 1866 at the encouragement of Wallace, Darwin embraced Spencer's now notorious phrase (see Marchant 1975, pp. 140–145). By the sixth edition of *Origin of Species,* natural selection and survival of the fittest had become synonymous, and the title of chapter IV was changed from "Natural Selection" to "Natural Selection; or the Survival of the Fittest." Darwin accepted Spencer's phrase "survival of the fittest" because it succinctly summarized what he expected from natural selection: fitter organisms surviving to increase in frequency over time. As previously discussed in chapters 2 and 3, "fitness" in this phrase means well designed. "Survival" means more than just life history survival; it means survival over evolutionary time, or evolutionary success. Thus, my translation of Darwin's phrase "survival of the fittest" is "evolutionary success of the best designed." Apart from use in this phrase, it is curious how rarely the term "fitness" appears in the *Origin of Species.*

"Survival of the fittest." Although overused in the popular literature and overanalyzed in the philosophical literature, this phrase is underappreciated by practitioners of evolutionary theory for the real issues it presents (see, for example, Endler 1986). The phrase "survival of the fittest" capsulizes the central tension between the notions of evolutionary success and good design that are inherent in the different uses of "fitness." These two basic senses of fitness are discussed further in the next chapter.

The phrase "survival of the fittest" has been seized on by critics of Darwinian evolution as reducing to an empty tautology. Who are the fittest? Those organisms that survive. If "survival" is the very criterion of "fitness," the principle of natural selection, as asserting "those organisms that survive are those that survive" is truly circular. So the tautology challenge runs, to the exasperation of evolutionary biologists who have little fear that their models of fitness relations reduce to empty truisms.

A definitive way to defuse the charge that "survival of the fittest" is analytic, true by stipulated definitions, is to show clear cases where it is false.

In phrases like "survival of the fittest," the word "survival" means not just life history survival, survival from birth to adulthood, but rather survival over the generations, or evolutionary success. In most models, this can be measured by a type's rate of increase. As we have seen repeatedly in this book, types with superior capacities often do not increase in frequency over time. So the phrase "survival of the fittest" cannot express a tautology, since tautologies cannot be false. Frequency-dependent fitness effects often frustrate the maximization of fitness at higher levels. The counterexamples to survival of the fittest encountered above can be classified into at least two alternative dynamical cases (Michod 1991), termed "survival of the first" or "survival of the common" and "survival of anybody."

I have previously considered how different selection paradigms may result from a mathematical representation of Darwin's conditions of heritability and variability in the struggle to survive and reproduce, including survival of the fittest, survival of the common (cost of rarity), and survival of anybody (cost of commonness) (Michod 1991). Based on Darwin's conditions, we should expect that the fittest (best designed) do not always win in evolution. Survival of the fittest is one of several possible paradigms for natural selection consistent with Darwin's principles. In alternative dynamical patterns, the fittest do not survive, even though Darwin's conditions for natural selection are met. As discussed in chapter 2, survival of the fittest (best designed) describes the outcome of natural selection only when reproduction is a linear function of density, or, equivalently, when population growth is exponential. In the "survival of the first" paradigm of natural selection (chapter 3), there is a cost of rarity in which new types cannot invade a population when they are rare, even if they are more adapted (have superior individual capacities). Survival of the first may help us to understand a very basic feature of the living world, the organization of life into relatively distinct species (see below). In addition, survival of the first characterizes the frequency-dependent advantage of hypercycles when they are established in populations at high density. In the "survival of anybody" pattern of evolution there is a cost of commonness, and types increase when rare independent of their abilities (see chapter 2, "Survival of Anybody," and chapter 4, "Population Dynamics and Natural Selection," for examples). A cost of commonness can result from standard ecological interactions that can trap populations at individual fitness minima (Abrams et al. 1993). In addition, a cost of commonness may apply to the first molecular replicators (Szathmáry 1991).

The theory of frequency- and density-dependent selection is replete with examples of natural selection in which the fittest genotypes and/or phenotypes do not increase in frequency; instead, natural selection may lead to a decrease in the average fitness of individuals and the population's demise. Other factors, including ecological interactions, the genetic system, linkage, pleiotropy, and the occurrence of within-organism change, can prevent the "fittest" organism—superior from the point of view of design—from being successful in evolution, that is, increasing in frequency over time.

Although it is easy to dispel the notion that the phrase "survival of the fittest" expresses a tautology, since "the fittest" don't always survive, it is another matter to clearly formulate uses of the term "fitness" in evolutionary explanation. I have already summarized the role of fitness in the emergence of new units of selection at the beginning of this chapter and how we may explain new levels of fitness by appealing to the dynamics of change in multilevel settings. The concept of fitness is of obvious concern to the adaptationist "survival of the fittest" pattern. But what role, if any, does the concept of fitness play in the other two patterns of natural selection, in which adaptive features (heritable capacities) do not predict evolutionary success, yet Darwin's conditions are met?

Heritable capacities and the underlying traits do enter into the dynamics of selection in these alternate paradigms; however, they are far from sufficient for predicting evolutionary success, especially when types must increase from rarity. If fitness was just construed as good design, then fitness could not help us to understand these alternate paradigms, even though Darwin's core conditions of natural selection are met. By taking a dynamical view of fitness as recommended by Fisher (1930) and by using the notion of $F$-fitness we can better understand these cases of natural selection (see, for example, figure 7-2 below and the accompanying discussion of Darwin's dilemmas).

## Surrogates for Natural Selection

Survival of the fittest thinking is just one of many shortcuts to understanding the outcome of natural selection. The complexity of the process of natural selection makes it only natural to search for shorthand descriptions. These surrogates typically focus on the expected products of natural selection and tend to ignore natural selection as a dynamical process of genetic change. Most selection surrogates assume that "fitness" is in some sense maximized by natural selection.

The adaptationist framework, based on survival of the fittest reasoning, remains prominent in evolutionary biology. Indeed, the view that the "fittest" should always increase in frequency has become so entrenched that a prevalent mode of thinking about natural selection, the optimality approach, consists of defining and maximizing individual fitness (or some function assumed to be closely related to individual fitness) (Parker and Maynard Smith 1990). Optimality theory assumes that more fit (better designed) traits increase in frequency over time. In other words, optimality assumes survival (increase in frequency) of the fittest, fittest in the sense of best designed.

As discussed further in chapter 4, optimization and maximization approaches to natural selection are often poor guides to the evolution of interactions when selection is frequency dependent. This is especially true in populations that are structured or when selection occurs on several different levels simultaneously. It is precisely this situation of multilevel selection that is most relevant to the emergence of new higher units of selection and the increase in complexity in evolution.

Nevertheless, maximization approaches to evolution are especially attractive as they may lead to simple principles of apparently broad generality. The principle of minimum entropy production advanced as a global description of adaptation (Glansdorff and Prigogine 1971; Prigogine 1980; Prigogine et al. 1972a,b) turns out to have dubious connections to the actual mechanics of natural selection (Bernstein et al. 1983). However, other maximization principles apply in certain cases. Under kin selection, the average inclusive fitness effect is maximized (see the section "Kin Selection" below). In the theory of selection in age-structured populations, the genotypic rate of increase (defined by a genotypic analogue of the Euler–Lotka equation) predicts many aspects of natural selection in density-independent populations, especially if selection is weak (see Charlesworth 1980, chap. 4, and Anderson and King 1970). These results have been used in the theory of life history evolution where workers typically solve for optimal life histories by maximizing the intrinsic rate of increase (Charlesworth 1980; Stearns 1992; Nur 1987, 1984; Stenseth 1984). In density-regulated populations, the genotypic carrying capacities determine the outcome of selection in equilibrium populations (Charlesworth 1980; Roughgarden 1979; Anderson 1971). Maximization principles have also been developed when selection results from standard ecological interactions between competing species (Leon and Charlesworth 1978). In populations that vary in time or space, maximization principles have been derived using different kinds of averages of individual fitness in the different habitats, such as the harmonic or geometric for variations in space or time, respectively (Levene 1953; Haldane and Jayakar 1963). There are many other examples.

Maximization principles are useful in that they lead to general principles for the evolution of phenotypes that may be gleaned from the selection dynamics of genotypes, even though these principles often apply only in ideal situations. It is important to remember their domain of limited application and that they are shorthand descriptions of a complex dynamical process.

There are also surrogates for fitness in the empirical realm, where biologists have sought and discovered correlates of fitness with varying degrees of success and generality. Heterozygosity and decreased fluctuating asymmetry are often, but not always, correlated with high fitness (fitness as measured by success). Heterozygosity may refer to protein heterozygosity or the lack of inbreeding. For bilaterally symmetrical organisms there is an assumed ideal state: perfect symmetry on the left and right planes. Fluctuating asymmetry (FA) refers to random left- and right-hand variations in a bilaterally symmetrical trait. There is an extensive literature beginning in the 1950s studying both heterozygosity and fluctuating asymmetry, and I do not intend to do it justice here. Instead, I provide a few references to introduce the topics (Lerner 1954; Møller and Pomiankowski 1993; Palmer and Strobeck 1986; Soule 1979). Quotation 23 in table B-1 below is typical of the view. The main question I want to consider is whether measures of protein heterozygosity or FA can be used as general measures of fitness. This use of symmetry is hinted at when Palmer and Strobeck (1986) say that "differences in FA among characters may provide a way of ranking traits in terms of their functional significance to an organism with little or no a priori knowledge of how those characters actually function."

Of course, correlation need not entail cause and effect. Indeed, workers have argued that the correlation of fitness (as measured by reproductive success) with both reduced fluctuating asymmetry and increased protein heterozygosity is mediated by a third factor, developmental stability in the face of stress. Developmental stability is often a "good thing," but periods of intense directional selection can result in a disruption of developmental stability with increased asymmetry. For example, during the incorporation of a new advantageous insecticide resistance mutant in Australian blowflies, developmental stability declined (Clarke and McKenzie 1987). Thus it is during periods of most intense adaptation that FA may increase. After incorporation of the advantageous mutation, developmental stability in the blowfly populations returned to its previously existing level. Consequently, after adaptation to a new external environment, FA again declines as modifiers begetting developmental stability evolve in the new genetic and physiological environment created by the advantageous mutation.

At best, heterozygosity and low FA are components of fitness in certain situations, but neither are sufficiently general, either in empirical support or in theoretical justification, to be considered as a general measure of fitness.

It would be remarkable, indeed it would be non-Darwinian, if it were possible for genetic properties alone, such as the degree of protein heterozygosity, to be a major determinant of fitness, since this would mean that the ecological context of selection was irrelevant. Today, many workers in population genetics seem focused on taking evolutionary problems down to the genetic level, ultimately to the DNA sequence itself, as if only in this way can we obtain unambiguous answers. I think the single most vexing problem in evolutionary biology is not the lack of molecular data, although this is a problem, but rather the lack of an adequate theory of fitness. It is fitness relations and selective transformations that make biology different from physics and chemistry.

## EVOLUTION OF SELFING

In chapter 4, we discussed how the average fitness of individuals in a population is often a poor guide to the outcome of natural selection, especially when interactions are important. This is especially true in the study of the evolution of reproductive systems. In this section I consider briefly the evolution of selfing and in the next section I consider the evolution of sex to see what we might learn from these classic problems in evolutionary biology about the meaning of fitness in evolutionary explanations. In the case of the evolution of selfing there is no increase in the average fitness of individuals in the population, no improvement in design, yet there is powerful selection nonetheless. In the case of the evolution of asexual females in a sexual population, all individuals are assumed to have the same individual fitness (expected reproductive success), yet asexual females double (approximately) in frequency every generation.

Fisher (1941) was dissatisfied with Wright's adaptive topography approach to evolution as it was often formulated in terms of average fitness of individuals in the population. Fisher felt (correctly) that maximization of average individual fitness was not a general result of evolutionary change. In chapter 4, we found that Wright generalized his adaptive topography concept to cases of frequency-dependent selection, in which cases the "fitness function" (equation 4-7) provides the basis for the topography instead of average individual fitness. Evidently, Fisher was unaware of this. Although dynamically sufficient (in its restricted realm), the "fitness function" bears no necessary relation to the average well-being of individuals in the population or group. In any event, Fisher wanted to provide an example of intense natural selection in which there was no effect on the average fitness of individuals in the population.

Fisher's example of the evolution of selfing assumes no differences in the survival and fertility (number of pollen grains and ovules produced) of selfers and outcrossers. Fisher also assumed that a selfer, in addition to fertilizing its own ovules, contributed pollen to the outcrossed pollen pool just like outcrossers. Although survival and fertility are assumed to be equal among the two kinds of plants, selfing increases rapidly in a population because of the greater representation of selfing alleles in selfed progeny. The selfer's ovules are not available for fertilization by pollen belonging to outcrossing plants, yet a selfer's pollen still has access to the ovules of these other plants. However, after the selfer increases and drives the outcrosser to extinction, there has been no improvement in the fitness of the individuals in the population. Fisher concluded that natural selection can be intense without there being any effect on the average fitness of individuals in the population.

> Natural selection, indeed a selection of great intensity, is at work, although by external criteria the average fitness of the species as a whole is entirely independent of the gene ratio. Individuals differ greatly in their fitness to the circumstances in which they are placed. The mean fitness of the species as a whole is, however, no guide to the selective intensities in action.

In Fisher's example of the evolution of selfing, there are differences in individual fitness during the increase of selfing, because selfers have greater mating success through pollen than do outcrossers (their fitness through female function, their number of ovules, is the same). However, after selfing goes to fixation there has been no increase in the average fitness of individuals in the population, that is, the total number of ovules fertilized after selfing evolves is the same as when the population was outcrossing. So there has been no improvement in the well-being of the population or of the individuals themselves; indeed, it would be difficult to maintain that the selfing population is better designed than the outcrossing one, even though there has been intense natural selection.

## Cost of Sex

In this section, I briefly consider one of the most fundamental problems in evolutionary biology, the evolution of sex, from the point of view of the role of fitness in evolutionary explanation. In the case of selfing just discussed,

Fisher observed that there can be intense natural selection without an increase in the well-being or fitness of individuals (after the evolution of selfing, individuals are no better off than they were before). A similar result can occur in the evolution of sexual reproduction in an asexual population, or in the evolution of diploidy in a haploid population, due to the masking of the deleterious effects of mutation (Bernstein et al. 1985a; Hopf et al. 1988). When outcrossing individuals arise in an asexual population, or when diploid individuals arise in a haploid population, outcrossers and diploids may enjoy a short-term advantage due to their greater masking of deleterious mutations. However, after the outcrossers or diploids are established, the mutation load must again increase to a new higher level. In the case of outcrossing sex, there is the additional burden of the costs of sex, and the overall fitness of individuals may be expected to decline as a result of natural selection.

The cost of sex involves the evolution of asexuality in a sexual population and provides an example of intense natural selection without differences in individual fitness (in the sense of the number of surviving offspring). I begin, following Maynard Smith, by considering two kinds of females that differ only with respect to whether their eggs are produced sexually or asexually (Maynard Smith 1978). All other aspects of their biology are assumed equal—the survival and fecundity (number of eggs produced) of the two kinds of females are identical. All sexually produced eggs are assumed to be fertilized. Males and females are also assumed to survive at the same rate. The only difference is that one-half of the eggs produced by sexual females develop into males, while all the eggs produced by asexual females develop into females. Although sexual and asexual females are assumed to be identical in individual fitness, asexual females double in frequency (approximately) each generation.

The argument is simplest in haploids. In this case there are just two genotypes, $g$ (sexual) and $G$ (asexual[2]). I assume equal sex ratio so the sexual genotypes are divided equally into males ($\male \male$) and females ($\female \female$). Since males and females are assumed to survive the same, the sex ratio never changes from unity in the sexual portion of the population. All $G$ genotypes are asexual females ($\female$). Every organism, whether male or female, sexual or asexual, survives the same with probability $s$, and all females produce the same number of eggs, $k$, whether they are sexual or asexual. In other words, the expected reproductive success of all the genotypes is assumed to be equal. Of course, there are differences between sexual and asexual females, but not in terms of their individual fitnesses; nevertheless, these other differences can lead to evolutionary change.

147

| | Initial | $g$ | $G$ |
|---|---|---|---|
| $g_{♀♀}$ | $\frac{1}{2}P$ | $\frac{1}{4}skP$ | 0 |
| $g_{♂♂}$ | $\frac{1}{2}P$ | $\frac{1}{4}skP$ | 0 |
| $G_♀$ | $R$ | 0 | $skR$ |

*Table 7-1*

Cost of sex in a haploid population. See tables A-5 and A-6 in appendix A for the diploid case.

The initial frequencies of the $g$ and $G$ alleles are assumed to be $P$ and $R$, respectively, as indicated in the first column of table 7-1 (with $P + R = 1$). The next two columns in the table give the relative frequencies of genotypes in the next generation among the offspring of the parents of the initial generation. In the case of sexual parents, only offspring of the same gender are included, so as not to count sexually produced offspring twice (females in the first row and males in the second row). Since the survival and fertility components of the individual fitnesses of all the genotypes are the same they cancel out, and the change in frequency of asexual females is $R/(R + 1/2P)$ $- R/(R + P)$, which is a doubling (approximately) in frequency when the asexual genotype $G$ is rare ($P >> R$). This gives rise to the so-called twofold cost of males (Maynard Smith 1978). Maynard Smith's model for the cost of sex involves between-species competition. However, we may expect parthenogenetic mutants to arise within sexual diploid populations, because this is known to occur in many species (*Drosophila* and other insects, fish, and lizards, to name a few). In appendix A, I show that a similar cost of sex exists when asexual mutants arise within a sexually reproducing Mendelian population of diploid individuals but not when the species is a self-fertile hermaphrodite.

Although individual fitness (expected reproductive success) is assumed to be equal for sexual and asexual females, the heritability of fitness is different, being twice as high in asexual females as in sexual females. We cannot see this in table 7-1 or table A-6, because there is no within-population variation in individual fitness (after all, these models were specifically constructed so that this was not the case). To calculate the heritability of fitness under the two modes of reproduction, we need a different model in which there is variability in fitness within both the sexual and asexual populations.

Heritability of fitness is another way of understanding the cost of sex, or the evolutionary advantage of asexuality. Even if we assume that the fitness

| Genotype | Parent Fitness | Genotype Frequency | Offspring Fitness Asexual | Offspring Fitness with Sexual Reproduction (random mating) |
|---|---|---|---|---|
| $AA$ | $W_{AA}$ | $P$ | $W_{AA}$ | $W_{AA,O} =$ $p'W_{AA} + (1-p')W_{Aa}$ |
| $Aa$ | $W_{Aa}$ | $2Q$ | $W_{Aa}$ | $\frac{1}{2}W_{AA,O} + \frac{1}{2}W_{aa,O}$ |
| $aa$ | $W_{aa}$ | $R$ | $W_{aa}$ | $W_{aa,O}$ $= (1-p')W_{aa} + p'W_{Aa}$ |

*Table 7-2*

Heritability of fitness under sexual and asexual reproduction. Under sexual and asexual reproduction, the heritability is one-half (assuming heterozygote intermediate) and unity, respectively. The fitness of the different genotypes is assumed to be the same under sexual and asexual reproduction. The frequency of the $A$ allele among gametes is $p' = (PW_{AA} + QW_{Aa})/(PW_{AA} + 2QW_{Aa} + RW_{aa})$. $W_{AA,O}$ and $W_{aa,O}$ are the average fitnesses of offspring of $AA$ and $aa$ parents, respectively, under sexual reproduction and assuming random mating.

of genotypes is the same under the two modes of reproduction, the heritability of fitness is different. Under asexual reproduction the heritability of fitness is unity while under sexual reproduction the heritability of fitness is 1/2 (assuming for simplicity in both cases no environmental variance).[3] This can be seen by calculating the regression of offspring fitness on parent fitness under the two modes of reproduction, using the appropriate columns in table 7-2 and the genotype frequencies at the start of the parental generation.

The twofold loss in the heritability of fitness under sexual reproduction applies regardless of dominance or whether there are separate sexes. Indeed, there are no males in the model in table 7-2; the model organism is assumed to be an outcrossing self-fertile hermaphrodite. Another way of expressing the twofold loss in heritability under sex is to say that the genetic relatedness (at all heterozygous loci) of offspring to parents is 1/2 under sex while it is unity with asexual cloning.

Although the heritability of fitness is lower in sexual populations, this does not mean that sex genes must decrease in frequency. It would mean this if fitness were really the same for sexually and asexually produced offspring, as figure 7-1 and table 7-2 both assume. There are many reasons to expect the fitness of sexually and asexually produced offspring to be different, including different levels of DNA repair, mutation load, and linkage disequilibrium, as discussed in three collections of papers on the topic (Michod and Levin 1988; Stearns 1987; American Genetics Society Symposium for the Evolution of Sex 1993). Be that as it may, the intrinsic advantage of parthenogenesis results from differences in the heritability of fitness, not differences in individual fitness (number of offspring).

## IMMORTALITY, DEATH, AND THE LIFE CYCLE

From parent to offspring, from parent cell to daughter cell, from DNA strand to daughter DNA strand, the cycle of life continues. Life does not begin anew each generation, but is passed on through time like a family heirloom. Most evidence supports the view that life began once around four billion years ago and has been passed down through the eons. Each of us can, in principle, trace our ancestry back in time to this ancient founder. Weismann referred to this immortal ancestry or cell lineage as the germ line.[4]

In 1890, Weismann first defined the immortal in biological terms (Weismann 1890). Weismann contrasted immortality with eternity; immortality is a state of activity and change in which the cycle of life continues indefinitely through time.

> And what is it, then, which is immortal? Clearly not the substance, but only a definite form of activity . . . the cycle of material which constitutes life returns even to the same point and can always begin anew, so long as the necessary external conditions are forthcoming . . . the cycle of life, i.e. of division, growth by assimilation and repeated division, should [n]ever end; and this characteristic it is which I have termed immortality. It is the only true immortality to be found in Nature—a pure biological conception, and one to be carefully distinguished from the eternity of dead, that is to say, unorganized, matter.

The continued well-being of life resides in the information encoded in genes, and life's immortal potential requires that genes be passed on in good repair. Sex functions to repair damaged genes and otherwise cope with genetic errors such as mutations (Michod 1995). In so doing, sex is an integral part of the well-being and immortality of life, in both unicellular and multicellular organisms.

In multicellular organisms, immortality requires totipotency. Sex, totipotency, and immortality became special characteristics of certain differentiated cells termed germ cells, in contrast to the somatic cells, which replicate by mitosis and are terminally differentiated into the various cell types that make up the tissues and organs of multicellular organisms. Why did immortality and reproduction become the function of one group of cells, the germ cells, with the other somatic cells becoming terminal? Again it was Weismann who first spoke convincingly on this subject.

According to Weismann, the germ-soma differentiation was invented in multicellular organisms because of the advantage to the fitness of the organism of specialization and division of labor among its cells. Somatic cells specialize at making bodies adapted to the contingencies of existence and germ cells specialize at making good gametes. Furthermore, once somatic cells began specializing in making bodies, they naturally would lose their immortality and capacity to divide forever. Why? Because unnecessary but costly structures or activities should be lost in evolution. If the germ cells specialize in immortality, there is no longer any need—from the point of view of the whole organism—for the soma to maintain this capacity. It all sounds reasonable, at least when you think in terms of the needs and fitness of the whole organism, and this was how Weismann approached the matter.

The problem with Weismann's approach is that, as we know from chapters 5 and 6, multicellular organisms did not always exist as evolutionary units, and, consequently, organismal needs were not always of paramount importance to selection. Before and during the transition from solitary cells to multicellular organisms, cells cannot be counted on to behave in the interests of the organism. Cells were, after all, evolutionary individuals in their own right for billions of years before the first multicellular organism emerged. Even now, with our individuality well protected by such marvelous adaptations as a germ line, immune system, and programmed cell death, humans are threatened by the evolutionary potential of extant microbes (witness the recent antibiotic-resistant forms of bacteria and other microbes).

Why would cells relinquish their evolutionary rights in the interests of the organism? There can be no question that division of labor among cells is important to the functioning of an organism; however, evolution must first settle the question of individuality—upon which evolutionary unit will selection focus, the cell, the multicellular organism, or some mixture of the two? We have clearly gotten ahead of ourselves by following Weismann's lead and depending on the needs of the organism to explain the ancient differentiation of the germ and the soma.

Why would cells relinquish their evolutionary rights in the interests of the organism? According to the studies reported in chapter 6 on the evolution of individuality (and reconsidered at the beginning of this chapter), conflict mediators arose during the course of multilevel selection of the organism and likely played a role in the origin of the ancient differentiation between the immortal (germ) and the mortal (soma) cells. By preserving the fitness gains at the level of the cell-group or organism (these fitness gains resulting from cooperation among cells), the germ line served to increase the heritability of fitness at the level of the multicellular organism and allowed it to emerge as an evolutionary individual.

Weismann's approach to the origin of the germ line clearly involved arguments based on individual, organism fitness (Weismann 1890). But what is the role of fitness in the explanation for the origin of the germ cells offered in chapter 6? According to this theory, the germ line is a feature that helps create fitness at the new emerging level by increasing its heritability. Organism fitness is not a cause in this theory; rather, organism fitness is an effect of the dynamics of change in a hierarchically structured population in which natural selection is occurring simultaneously on several different levels at once. By improving the covariance between fitness at the new organism level and the genes carried in the zygote state, germ line modifiers increased in the population. It is the problem of the heritability of fitness at the new organism level that likely shaped the evolution of the germ line.

## Kin Selection of Altruism

There is probably no clearer example of the tension between organism fitness and the dynamical approach to fitness ($F$-fitness, introduced in chapter 1) than when selection acts on behavioral traits in populations in which kin are present. I used kin selection in chapter 3 to study the origin of cooperative networks of genes as a precursor to the first cell and again in chapters 5 and 6 to study the transition from cells to multicellular organisms. The theory of kin selection and inclusive fitness was presented briefly in chapter 4. I now consider the kin selection explanation of altruism in a more philosophical manner for the lessons it teaches about the role of fitness in evolutionary explanation.

In the simplest case, there are assumed to be two heritable capacities for each genotype. These heritable capacities summarize the way in which the different kinds of individuals (genes, cells, or organisms) behave in terms of the effect of the individual's behavior on its individual fitness (expected reproductive success) and the fitness of its partner: $c_i$ is the additive effect of genotype $i$ on its own survival, and $b_i$ is the additive effect of genotype $i$ on its partner's survival. Different kinds of behaviors are considered by fur-

ther constraining $c_i$ and $b_i$. A particular case of interest is an altruistic behavior, defined as a costly cooperative behavior that decreases the survival
of individuals exhibiting the behavior while increasing the survival of individuals receiving the effects of the behavior. Consequently, for altruistic behaviors, $c_i < 0$ and $b_i > 0$. By behaving altruistically, individual fitness is
decreased by $c_i$. Examples of altruistic behaviors include the sterile castes
in the social insects, replicators making proteins early in the history of life
(chapter 3), and cooperation among cells in a multicellular organism (chapters 5 and 6). To complete the model, for our purposes here, we need only
specify the environment.[5] The aspect of the environment that is of interest
is the kinship structure in the population within which the behavioral interactions occur, represented by $R$ in the inclusive fitness approach given in
equations 4-20, 4-21, and 4-22. For this additive fitness model, individual
genotypic fitness defined in equation 4-17 becomes

$$W_i = \sum_j (1 + b_j)P(j/i) + c_i. \tag{7-1}$$

In equation 7-1, $P(j/i)$ is the conditional probability that an individual of
genotype $i$ interacts with an individual of genotype $j$, and $j$ ranges over all
genotypes present in the population. These conditional probabilities are
themselves averages over different kinship contexts and can be calculated
from the population structure or kinship structure as shown in Michod
(1982). The point here is that the individual fitness of a particular genotype
$i$ is not a property of $i$ alone, but is rather a frequency-dependent property
of all the other genotypes $j$, through their behavior, $b_j$, weighted by the probability of interacting with them, $P(j/i)$.

Furthermore, the conditional probabilities $P(j/i)$ are themselves averages
over the different kinship contexts, family units in the simplest case of selection in a family-structured population (see Michod 1982 for a review).
Individual genotypic fitness is a property of the entire population in a frequency-dependent manner. For these reasons, no particular individual has
fitness $W_i$ (except, perhaps by chance), since $W_i$ is a statistical construct.
For example, in family-structured populations (behavior between full siblings), any particular individual organism is found in only one family type
(there are nine family types in a standard one-locus model) and its individual fitness depends only on its family context, yet the average given in equation 7-1, $W_i$, is a property of all family types.[7] Such a construct cannot be
a measure of adaptedness of an organism, or of an organism's propensity
for fitness (see chapter 8), since it is not a property of the organism at all.

153

Although altruism is costly to the expected reproductive success of the altruist, the genotypic fitness of altruists defined in equation 7-1 can exceed that of nonaltruists and altruism can increase in the population. There is no paradox here. In any specified context, nonaltruists survive and reproduce better than altruists, yet when averaged over all contexts, altruists may do better.

Kin selection illustrates well the problems inherent in using an organism's individual fitness as a measure of its overall adaptedness or design. Is the adaptedness of a sterile worker bee zero because it has no offspring? Clearly not, the worker is well adapted to its function and its environment. It is just that when relatives are present in the environment, the effect of an organism's behavior on its own survival and reproduction is insufficient to understand natural selection, even if that effect is to be sterile. The condition for natural selection in this case (recall equation 4-21) involves the environment, $R$. Consequently, there is no simple measure of an organism's overall adaptedness (that is a property of the organism alone) that does the job of predicting the outcome of natural selection.

The only way in which adaptedness or good design could be said to enter into kin selection is through individual properties, the heritable capacities, $c_i$ and $b_i$. These capacities summarize the properties of individuals that are relevant to the explanation at hand (for example, sterility and helping behavior in the case of sterile insect castes). There is no useful notion of overall adaptedness, no way of combining $c_i$ and $b_i$ into a measure of good design that is a property of the organism. There are just the heritable capacities themselves, the traits upon which they depend, and the way in which these individual capacities enter into the process of natural selection as explicitly specified in the dynamics (equations 4-17, 4-21, 4-22, and 7-1).

## HETEROZYGOTE SUPERIORITY

Heterozygote superiority is one of the most worn examples in elementary population genetics, yet it has a lesson to teach us about fitness. It illustrates why "survival of the fittest," fittest in the sense of individuals best adapted to survive and reproduce in an environment, can be false—if there is sex.[8] Consider a single gene locus in a random-mating, diploid organism, in which heterozygotes (*Aa*) are assumed to be the most adapted, and, as a result heterozygotes have a greater likelihood of surviving to reproduce than individuals that are homozygous (*AA* or *aa*). Let $W_i$ be the expected reproductive success of individuals of type $i$ (individual genotypic fitness), and $X_i$ the frequency of type $i$; fitness is assumed constant for simplicity ($i = 1$,

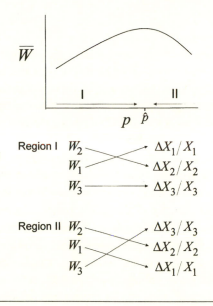

*Figure 7-1*

Individual fitness and evolutionary success with heterozygote superiority. Under heterozygote superiority, $W_2 > W_1 > W_3$, but the heterozygote never has the highest per capita rate of change. The equilibrium gene frequency is $\hat{p}$. Adapted from Michod 1995.

2, and 3 stands for genotypes *AA, Aa,* and *aa,* respectively). I ask how the genotypic rate of increase over time, fitness in the sense of Fisher (see equation 1-1), depends on $W_i$. Fisherian fitness is defined as the per capita rate of increase; the change in frequency after a single generation is $\Delta X = X_{t+1} - X_t$, so the per capita rate of increase is $\Delta X/X$.

Figure 7-1 shows the relationship between individual fitness, the $W$'s, and the evolutionary success as measured by the per capita rate of increase. This relationship is different at different stages of the evolutionary process (below or above the equilibrium gene frequency $\hat{p}$), but at no stage does the heterozygote have the greatest per capita rate of increase, even though heterozygous individuals are most adapted at surviving to reproduce. "Survival of the fittest" predicts the same ranking of evolutionary success and fitness. If $W_2 > W_1 > W_3$, then based on survival of the fittest reasoning we should have $\Delta X_2/X_2 > \Delta X_1/X_1 > \Delta X_3/X_3$. However, this never happens. Because of sex the heterozygous genotype cannot breed true and so segregates out the less fit homozygotes. As a result, the homozygous genotype that is currently underrepresented has the highest per capita rate of increase.

155

## SICKLE CELL ANEMIA

An example involving heterozygote superiority is the evolution of sickle cell anemia in human populations, one of the touchstone cases of natural selection. There are many treatments of this problem; my presentation follows closely that of Templeton (1982). The sickle cell trait is determined by a single allele at the β-chain hemoglobin locus. The sickle cell allele reduces the capacity of hemoglobin to transport oxygen. The sickle cell condition is common in West and Central Africa, where the sickle cell gene reaches frequencies of 16% and approximately 5% of the population may die in infancy. The frequency of this gene among United States Blacks is about 5%. The geographic distribution of the trait is positively correlated with malaria. Three alleles have been described at the β-chain hemoglobin locus, the wild-type allele, $A$, in frequency $p$, the sickle cell allele, $S$, in frequency $q$, and a third recessive allele denoted $C$, in frequency $r$. Survivorship to adulthood data for the diploid genotypes in one West African village are given in table 7-3.

We would like to know why the sickle cell gene is so common in West African populations. Furthermore, why doesn't the $C$ allele increase, as it has the highest individual fitness and would seem to represent the optimal solution?

The answers to these questions are obtained by studying the gene frequency equations for the sickle cell allele, $S$, and the recessive allele, $C$. This is done in Templeton (1982) and will not be repeated here as the basic points can be made without the algebra. The first point to recognize is that human populations are sexual and heterozygotes cannot breed true. With regard to

| Genotype | Survival | Condition |
|----------|----------|-----------|
| AA | $W_{AA} = 0.9$ | malarial susceptible |
| AS | $W_{AS} = 1.0$ | malarial resistance |
| SS | $W_{SS} = 0.2$ | anemia |
| AC | $W_{AC} = 0.9$ | malarial susceptible |
| SC | $W_{SC} = 0.7$ | anemia |
| CC | $W_{CC} = 1.3$ | malarial resistance |

*Table 7-3*

Sickle cell anemia. The struggle to survive in one West African village. Survivorship expressed relative to AS genotype assuming a malarial environment.

the $A/S$ competition, the sickle allele is maintained in the population by the advantage of the $AS$ heterozygote. But why doesn't the $CC$ genotype increase and predominate in the population?

The answer, according to Templeton (1982), lies in the initial conditions before the spread of malaria, and the fact that the $C$ allele is recessive. Malaria began spreading with the advent of slash and burn agriculture (which created breeding sites for mosquitoes) of the Bantu-speaking peoples around 200 B.C. Without malaria as a selective agent, the $A$ allele is favored ($AA$ has highest survival), and so we assume the initial population was fixed for the wild-type $A$ allele with rare mutations to $S$ and $C$ occurring in the population. When malaria became a selective agent, the $S$ and $C$ mutations were carried in the heterozygous state, because rare mutations are almost always found in the heterozygous state. This allowed the $S$ allele to begin increasing, as its relative survival in heterozygotes in a malarial environment is (table 7-3) $W_{AS} = 1.0 > W_{AA} = 0.9$. The $S$ allele increases even though homozygous $SS$ genotypes will begin appearing when the $S$ allele reaches appreciable frequency in the population. Natural selection is short-sighted. The $C$ allele is neutral when rare, as its survival in the heterozygous state is $W_{AC} = 0.9 = W_{AA} = 0.9$. Even though $CC$ homozygotes are most adapted, natural selection doesn't lead there. Instead it produces a far-from-optimal outcome with much suffering in the affected human populations.

The explanation for sickle cell anemia given here (and in Templeton 1982) involves several components in addition to considerations of individual fitness (as a probability of survival): there is sex, the genetic system is diploid and heterozygous individuals have higher fitness in a malarial environment, the mating system is assumed to be random mating (otherwise homozygous genotypes could exist in appreciable frequencies and $CC$ genotypes might have increased initially), rare mutations are found in the heterozygous state, and the initial state of the population was free of malaria and so the wild-type allele predominated initially.

In sum, the sickle cell example teaches that the outcome of natural selection depends upon individual capacities that are heritable, the environment, genetic system, reproductive system, mating system, and population structure. Natural selection does not create the best or optimal solution (in the case of sickle cell this would be a pure $AS$ or $CC$ population). Natural selection, and this was Templeton's (1982) main point, is a complex dynamical process where the consequences are not determined simply by individual fitness (see table B-1, quotation 26).

## DARWIN'S DILEMMAS

In *The Origin of Species* Darwin asked, "Why is not all nature in confusion instead of the species being, as we see, well defined? . . . Why are not all organic beings blended together in an inextricable chaos? . . ." In other words, why do species exist as relatively distinct groups? This was the central problem Darwin wanted to solve, as advertised in the title of his great book. Darwin's attempt to answer the species question led to his two dilemmas: missing links in habitat space and missing links in time. As background, let me recount Darwin's logic from *The Origin of Species.*

"As according to the theory of natural selection an interminable number of intermediate forms must have existed, linking together all the species in each group by gradations as fine as are our existing varieties, it may be asked; Why do we not see these linking forms all around us? . . . Why, if species have descended from other species by insensibly fine gradations, do we not see innumerable transitional forms?" This is his dilemma of missing links in habitat (or morphological) space among current living forms.

His answer to this problem was based on his view that natural selection should produce "innumerable transitional links" if evolution occurred in "an extensive and continuous area with graduated physical conditions." He argued that intermediate types are not observed among current living forms because the intermediate regions between two species are small. Small regions mean small population sizes, and he believed that small populations evolve more slowly than large populations and therefore would be eliminated in the long run by competition with the faster-evolving species. In his words, "the neutral territory between the two representative species is generally narrow in comparison with the territory proper to each. The intermediate variety, consequently, will exist in lesser numbers and during the process of further modification through natural selection, they will almost certainly be beaten and supplanted by the forms which they connect; for these from existing in greater numbers will, in the aggregate, present more variation, and thus be further improved through natural selection and gain further advantages."

Darwin realized that his solution to the dilemma of missing links in habitat space created his second dilemma of missing links in time since "just in proportion as this process of extermination has acted on an enormous scale, so must the number of intermediate varieties, which have formerly existed, be truly enormous." This is his dilemma of missing links in time. As is well

158

known, to resolve the second dilemma he appealed to what he viewed as "the extreme imperfection of the geological record."

Why do species exist? I now consider a hypothesis proposed by my colleagues and I (Bernstein et al. 1985c; Hopf and Hopf 1985; see also Michod 1995), from the point of view of the role of fitness in evolutionary explanation. Our main point was that there is a fitness cost of rarity in sexual species, which stems from the nonlinear density-dependent terms implied by mating. The cost of rarity may be interpreted either as a cost of finding an appropriate mate, or as the cost in rare species of allocating time, resources, and energy from the other components of fitness so as to insure mating. As discussed in more detail in the aforementioned sources, this fitness cost destabilizes a continuum of closely packed species, each slightly different from one another (Darwin's "nature in confusion"), and creates in its place a stable set of relatively distinct and discrete groupings. There may be other reasons why species are distinct; however, the fitness cost of rarity implied by sex is sufficient, as has been confirmed by recent mathematical modeling of this hypothesis (Noest 1997).

The general evolutionary consequences of a cost of rarity are best interpreted by using fitness in Fisher's sense, $F(y)$ (figure 7-2). The abscissa is a continuous resource or environmental parameter, $y$, such as seed size or height up a mountain. The variable $y$ may also be interpreted as the value of the trait utilizing the resource or adapting to the environmental parameter. The optimum for the population is $y = s$. The ordinate is the per capita rate of increase, fitness in Fisher's sense, as a function of $y$, $F(y)$. The current state of adaptation of the population is represented by $y = y_0$. I assume that the initial population has reached equilibrium in its environment so that it is neither increasing nor decreasing in numbers.

In the top panel of figure 7-2 is illustrated survival of the fittest. Any new type with a trait value closer to the optimum has a positive rate of increase and so the population evolves toward its optimum. In the middle and bottom panels is illustrated the cost of rarity, represented by the spike in fitness for the resident population (advantage of commonness), or, equivalently, the suppression in fitness of nonresident types. Using Darwin's language we describe this as "survival of the first," because the resident type suppresses the fitness of other types if they must increase from rarity. The middle panel shows the case where the resident population is stable to all new trait values; a kind of stasis has been attained regardless of the adaptedness of invading types. For this reason, a central premise of Burns' analysis of fitness (Burns 1992), that stasis reflects adaptedness, may be without theoretical

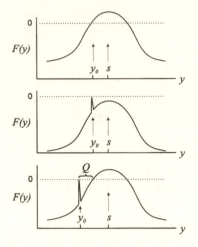

*Figure 7-2*

Consequences of cost of rarity. Fisherian fitness is plotted as a function of trait *y*. Redrawn from Bernstein et al. 1985c.

justification. In the bottom panel is the case where the population is poorly adapted and new types may increase if they differ from the resident by quantum *Q*, which represents a limit to similarity; in other words, distinctness of types called species.

Having considered the role of fitness in a wide range of evolutionary explanations, from the major transitions in evolution to specific cases of adaptation, I propose in the next chapter a general framework for the role of fitness in evolutionary explanation.

# A Philosophy of Fitness

## DYNAMICS OF DESIGN

Not so long ago, before Darwin, the source of design evidenced in the natural world was found outside nature in the transcendent will of a Creator. Where is the source of design to be found today, now that we look within the workings of life to explain the forms life takes? I accept the position of most working biologists that biology need presuppose nothing beyond physical and chemical mechanism, natural selection, and history (Williams 1992). We are coming to understand more about how natural selection works and how design emerges from the "pathless wastes of generation and transformation" (Dewey 1909), as organisms and other units of selection busily prepare for (and often fail to meet) the next of life's many contingencies. Design in nature is created by adjustment of a population's gene and genotype frequencies through trial and error—by natural selection rather than creation according to premeditated design.

Design in biology does not come from a preexisting plan, rather it is created through differential replication and survival operating on random variation. In other words, design is created by dynamics, the Darwinian kind of dynamics that occurs during natural selection. Can there be a meaningful sense of design without a preexisting plan? I think the answer to this question must be yes. Indeed, I wonder whether there is ever really a plan, even in human thought. Let me explain. It is often observed that one of the most efficient ways of solving a design problem in engineering is by using a Darwinian solver based on the principles of natural selection: randomly generated solutions with differential amplification based on a measure of performance of the variant (Holland 1992). Why do these Darwinian solutions often mimic their rational counterparts? The same question may be asked in another context: Why do organisms often behave in optimal ways? I think the answer to these questions may be that the human mind is also a Darwinian system (Edelman 1987). For an explanation of the principles of neural Darwinism oriented to evolutionary biologists, see Michod (1989, 1990). If human thought is a Darwinian system, could it be that design in any context requires, not a plan, but the dynamics of selection based on a system organized using Darwin's principles of variation and differential fitness? It is commonplace to believe that the existence of design requires a plan and a planner. Ultimately, only the plan of a divine Creator can explain the design inherent in the living world; so argued William Payley in 1802. Darwin undercut that argument for premeditated design and the existence of a divine Creator. Will Darwin's principles also undercut the argument for existence of premeditated plans in human thought? Could it be that the existence of design, in all circumstances, requires no plan, but just Darwinian dynamics?

The repetitiveness of the life cycle along with the predictability of the environment allow the products of selection during past contingencies to appear as design in the present. The remarkable fit of these activities to the contingencies of existence makes living possible. As just mentioned, such adaptive design has been used, in various arguments from design, as incontrovertible evidence for the existence of a divine Creator. Today, we take these same designs as evidence of the power of natural selection, or in the terminology used in this book, as evidence of the central role of heritable capacities in determining evolutionary success.

How is all this adaptation, and fitting of means to ends, possible? And where do organisms come from? I have tried to understand the origin of multicellular organisms in terms of the principles of cooperation and conflict in a multilevel selection context. Organisms routinely develop structures that are needed later in life. To the extent that programs for these structures are encoded in the genes, how can the genes "know" beforehand what the organism will need? It is as if the arrow of time has somehow been reversed; with the end directing the means; with the future reaching back to affect the present. Of course, it does not make sense to say that future goals cause present events, so how are design and purpose (the teleology so magnificently exhibited in the forms of life) possible? Natural selection, the agency of design, makes purpose possible.

Life involves cycles that repeat in time. The seed germinates, develops into the adult form, and through birth gives rise to more seeds, similar in constitution, all in an environment that is to some degree predictable. The structures and behaviors active in the present exist in part because of their contributions to the fitness of ancestors living in the past. There are other reasons that structures and behaviors exist, but natural selection based on differences in fitness is the primary directive and creative force in evolution. Living systems are maintained by fitness relations not just in time and space but also in organization.

In certain organizations, the tendency of living entities to increase their fitness must be rigidly controlled. To build ever more complex multicellular organisms or social organizations, lower-level units of selection must, as it were, relinquish their claim to flourish and multiply—their access to fitness must be constrained. This occurs whenever there is a passage to a new unit of selection, as has occurred several times in evolution, for example, when genes form networks of cooperating genes (chapter 3), when groups of cells form multicellular organisms (see chapters 5 and 6), or when organisms cooperate to form societies.

At any level in the biological hierarchy, the cycles of life are nested, and lower-level replicators (genes, chromosomes, cells, and tissues) often maintain themselves by behaving in the interests of the higher level, for example, the multicellular organism. But their partnership is never complete— we refer to misbehaving lower-level units as diseases and parasites. Adaptive design for the organism depends upon the regulation of these lower-level units, and this regulation allows the organism to emerge as a unit of selection, or an individual, in its own right. In this way, purpose becomes possible for the organism. I have tried to explain how this occurs.

## WHAT MAKES BIOLOGY DIFFERENT?

All sciences are concerned with transformations of one form or another. Witness the dramatic transformations of everyday experience: the storm clouds build during the afternoon, the match ignites in flame. What, if anything, distinguishes biological transformations from the transformations occurring in the nonliving world? Biological units—genes, chromosomes, cells, organisms—are made up of physical and chemical compounds, and, of course, must obey physical laws. Biological transformations are also purposeful—each stage prepares the way for the next, each structure and behavior prepares the organism for life's many contingencies. What is the source of this purpose?

Biological transformations are driven by the dynamics of self-replication. The life cycle of the organism depends upon replication of its component cells. The replication of cells depends upon the replication of internal structures and chromosomes. The replication of chromosomes depends upon the replication of genes and on the complementary base pairing inherent in DNA. Complementary base pairing is a special property of nucleic acids and depends upon electrostatic interactions between the nucleotide bases. Consequently, replication and reproduction at all levels, including the life cycle of the cell and the organism, depend upon electrostatic interactions, that is, upon physical-chemical properties. What is it that is uniquely biological about these processes?

It is the role of differential replication and natural selection as the agents of design that makes biology different from other sciences. I discussed the general properties of natural selection and how it differs from commonplace selection in chapter 1 and in chapter 4 in the section "Fisher's Fundamental Theorem." In any selection process, there is a bias in the selected set according to the properties the individuals possess. In biological selection, these properties (termed heritable capacities in this book) cause differences in fitness. There are other causes of bias, as we have seen; however, fitness relations based in part on heritable capacities mediate biological transformations at all levels (recall, for example, equation 1-5). Everything else in biology is molecules, the dynamical outcome of chemistry and physics, or a remnant of history. This is what I understand Lewontin to mean when he says, "Natural selection of the character states themselves is the essence of Darwinism. All else is molecular biology." (Lewontin 1972).

Biological transformations are directed towards ends. Physical transformations are also directed towards end states, often states in which matter is disorganized and entropy has increased. However, the end state of physical change plays no special role in starting the transformation anew; ends don't create beginnings in the physical world. For this reason there is no meaningful way to speak of the fire or storm as designed for some purpose. In biology, matters are altogether different—ends do create beginnings (the seed germinates to develop into the adult, which produces more seeds) because of the repetitiveness of life cycles, and for this reason purpose is possible.

## SUCCESS AND DESIGN

Two basic distinctions in the meaning of "fitness" emerge from the cases of natural selection studied here and from the commonly used fitness phrases given in table B-1 in appendix B. First, there is fitness as operationally measured by success. For example, the standard definition of individual fitness is the expected number of gametes produced. Quotation 6 by Endler in table B-1 is an especially explicit example of fitness as a measure of success (see also his sense 1 of "fitness" in Endler 1986, p. 40, table 2.1).[1] Another example of operational fitness is fitness in Fisher's sense as measured by the per capita rate of increase of a type—the left-hand side of $F$-fitness equations. There are many other examples of operational fitness. Until these success-based measures of fitness are decomposed into their components and reconstructed in the context of the environment and the genetic and reproductive system, there is little opportunity for real explanation, and their use in evolutionary explanations is likely to be tautological.

We want to know *why* a type is successful at survival, at reproduction, or at increasing in frequency. This gives rise to the second sense of fitness: the properties and capacities of the individual, often the organism, that causally contribute to success. Sometimes "good design" or "adaptedness" are used to refer to these causal components of success. Evolutionary ecology is devoted to understanding the design components of success (phenotypic capacities that contribute to success) and to answering the questions, "What is good design in this environment?" and "For what purpose is this structure or behavior designed?" As transitions occur in the units of selection, good design must shift its reference from the lower-level to the new higher-level unit, although the shift is never really complete. In the models studied here, *heritable capacities* refer to the individual capacities and properties of the phenotype that causally contribute to success (and are passed on from parent to offspring).

Let's take the far simpler example of whether a key fits a lock. First, we could try the key and see if it unlocks the door. This is the criterion of operational success. But having decided that the key fits the lock, we are in no better position to know what it is about the key that allows it to open the lock. Is it the shape of the key or is it a special magnetic code programmed into the key? Second, we could measure the shape of the key and the shape of the tumbler in the lock and predict, a priori, whether the key will fit the lock. This is analogous to measuring the fitness of the key in the sense of good design, the design of the key to fit the lock.

The essence of Darwin's phrase "survival of the fittest" discussed in the last chapter lies in the distinction between adaptive attributes of the individual and evolutionary success. Darwin's phrase means that survival (fitness in the operational sense of success) can be explained by fitness (fitness in the sense of good design). The distinction between fitness as success and fitness as design is also the basis for the "two ways to find out about the fitnesses of traits in a population" proposed by Mills and Beatty (1979): either look at the expected consequences of high fitness (increase in frequency) or investigate the expected underlying cause (phenotypic attributes of the individuals involved). This same distinction is behind Sober's (1984) discussion of "selection for" (the underlying cause) and "selection of" (the consequences such as a change in frequency) and in Brandon's treatment of fitness in chapter 1 of his book (Brandon 1991).

In the studies in the present book, we have found that design attributes of

165

the individual (what I have termed heritable capacities) are one of several reasons why traits increase or decrease in frequency during the process of natural selection. To say that design is one of several components of a type's rate of increase is not to say that design attributes are unimportant to evolutionary success, but rather to acknowledge the inherent complexity of natural selection among hierarchically structured units.

## LONG-TERM VERSUS SHORT-TERM MEASURES OF FITNESS

I have not considered long-term measures of fitness, because natural selection primarily operates in the short term; it is opportunistic and short sighted, as I have mentioned several times in the preceding chapters. Although measures of fitness have been proposed for long-term success, special difficulties arise in expressing long-term evolutionary success in terms of heritable capacities of the units under study in a noncircular way (see also Endler 1986, pp. 36–38). This is especially true when the units are higher taxonomic categories. In considerations of long-term fitness, the unit of selection may be the gene, as in Fisher's branching-processes calculation of the probability of survival of a new advantageous mutation (Fisher 1958), or in Cooper's expected time to extinction of an allele (Cooper 1984). Often the focus is on a higher evolutionary unit such as a population, species, or higher taxonomic level (Cooper 1984; Nur 1987, 1984; Thoday 1953).

I have also not considered the meaning of fitness in a time-varying environment. Fitness may vary in time in a deterministic fashion (Haldane and Jayakar 1963) or stochastically (Gillespie 1973, 1974, 1977), and different measures of evolutionary success are relevant in each situation. I believe that the general approach proposed here (of relating the relevant measure of evolutionary success to heritable capacities of the individual units along with aspects of the genetic system and environment) can be extended to time-varying environments, even though I have not done so.

Burns (1992) provides a framework for relating long- and short-term measures of fitness in evolutionary biology. Although he states, "Fitness is a consequence of the adaptedness of an entity to its environment" or that the point of his "expanded concept of fitness is to capture how adaptedness can be manifested as stasis," at no point in his analysis does he make this relationship explicit. All references to "adapt" words in his analysis are superfluous. His initial approach is similar to that taken here in that he defines evolutionary entities in terms of their heritable attributes and is interested in mapping evolutionary trajectories in these entities through time. Starting with a species, descendents are identified along with their attribute values and divergence and persistence times. He adopts a success-based definition of fitness (see table B-1, quotations 24 and 25) that takes into account the similarity of descendants (in the space of attributes) and their persistence times. However, he does not express this success as an $F$-fitness–like function of the attributes he defines. For example, in his examination of tooth dimensions in fossil horses, we never know why it is that the particular attributes studied cause different species of horse to have different fitnesses (even though fitness is defined in terms of these attributes).

This is not to say that success-based definitions of fitness at higher taxonomic levels cannot be made noncircular. For example, it may be the case that larval properties of marine invertebrates affect speciation and that a model expressing speciation as a function of larval properties and the environment could be used to study the persistence and extinction of different marine invertebrate clades (Jablonski 1986, 1997).

A comparable example in micro-evolution involves calculating fitness from changes in trait frequencies, as is often done by fitting selection models to gene and genotype frequency data (Prout 1971a,b, 1969, 1965; Du Mouchel and Anderson 1968). The fitted fitnesses obtained in this way can be used to measure the strength of selection or to evaluate the suitability of a specific genetic model. However, having estimated the fitnesses, we don't know why the gene frequency changes occurred; the fitted fitnesses do not explain the changes in frequency in the sense of relating the observed changes to the underlying traits and ecology. Nevertheless, such fitness estimates are useful in providing a starting point for deconstructing fitness into a causal model of individual attributes and the environment (taking into account the genetic and reproductive system).

## Darwinian Dynamics

Let us now reconsider the general framework proposed here as a context for investigating the meaning of fitness in evolutionary biology. Mathematical models in population biology fill out and make explicit Darwin's general framework for natural selection as expressed in his conditions of variation, heritability, and the struggle to survive introduced in chapter 1, "Darwinian Dynamics." The goal of these models is to explain evolutionary change and design on the basis of natural selection.

Many different examples have been considered in this book.[2] We used replicator dynamics in chapter 2 to study the evolution of more adapted molecules and in chapter 3 to study protein replication and the emergence of cooperative networks of replicators termed hypercycles, culminating in the creation of the first true individual, the cell. In chapter 4, we developed the basic gene frequency equations of population genetics in the context of the evolution of interactions between members of the same species. We extended these basic equations in chapter 5 to study the dynamics of cooperation among cells within organisms and, in chapter 6, to study the evolution of modifiers of within-organism change so as to provide a theoretical framework for the evolution of individuality and transitions in the units of selection. Chapters 5 and 6 were concerned with the emergence of a new unit of selection, the multicellular organism, from groups of cooperating cells. Cooperation was the underlying trait that mediated fitness relations at both the cell and the cell-group or organism level. In chapter 6, I measured the progress towards individuality at the new organism level by studying the effect of conflict mediation on the heritability of fitness at the new organism level. Other examples of the dynamical approach to evolutionary problems involving natural selection were discussed in chapter 7.

The dynamical equations underlying natural selection are examples of a general schema for the process of natural selection, termed the "Darwinian dynamic" by Bernstein et al. (1983). Darwinian dynamics are systems of equations that satisfy Darwin's conditions of variability, heritability, and the struggle to survive and reproduce. These equations exhibit a common structure, and by considering this structure, we may clarify philosophical questions about the nature of evolutionary explanation and the role of fitness, since the models make explicit distinctions that are difficult to express verbally (Bernstein et al. 1983; Byerly and Michod 1991a,b; Michod 1984, 1995, 1986, 1981).

Consider one of the simplest examples of a Darwinian dynamic, $\dot{X}_i = X_i(b_i - d_i - \Psi)$ (Eigen's form of template-mediated replication studied in equation 3-2), in which the per capita rate of increase of replicator $i$ is expressed as a function of the density of type $i$ in the population, $X_i$, the genotype-specific capacities for birth, $b_i$, and death, $d_i$, and the resources available in the environment.[3] I have referred to the capacities at birth and avoiding death (and similar genotype-specific capacities) as "heritable capacities" because they represent design properties of the genotype and satisfy Darwin's condition of heritability.[4] Resources represent the limiting factors in the environment and underlie the "struggle to survive" in the simple model considered in equation 3-2.

Although there are general conditions that underlie all cases of natural selection (Darwin's conditions), there is no single equation that can represent all cases. The form of the equation used to model natural selection and the particular capacities posited grow out of the question being asked. This is because reproduction takes different forms for different evolutionary units, and the "struggle to survive" has different meanings in different situations. The "struggle" may not even involve survival in a life history sense, but mating or reproductive success. In chapters 5 and 6, I studied the consequences of within-organism change for the emergence of higher levels of selection. The struggle to survive was represented directly in terms of fitness effects at the different levels. The underlying traits and causes of these fitness effects were not explicitly modeled, as is common in models of social behavior in which fitness effects are traded between individuals. The heritable capacities in our studies of individuality involved the costs and benefits of cooperation and rates of mutation in the different cell lineages.

## NATURAL SELECTION AS A BIOLOGICAL LAW

The interpretation of the variables and parameters in Darwinian dynamics provides the basis for an ontology of entities involved in natural selection (Byerly and Michod 1991a; Bernstein et al. 1983; see also Endler 1986). Using Darwin's conditions, given a localized population of individuals there must be:

*Variation.* Different types indexed here by the subscript $i$ (each in density $X_i$) have different traits and capacities. In the case of genetic selection, $i$ stands for the genotype class within the population.

*Fitness.* Evolutionary biologists are interested in the evolution of traits affecting fitness. The "struggle for existence" has as many different meanings as there are causes of fitness differences. Nonlinearities in the dynamics of population growth prevent types from increasing without bounds, an idea Darwin borrowed from Malthus. In the replicator dynamics of chapters 2 and 3, I represented the struggle for existence in terms of the competition of genotypes for finite resources, $R$, that were necessary for reproduction. In addition to resource competition, fitness interactions contributed to selection among the molecular replicators studied in chapters 2 and 3. In chapters 5 and 6, I studied the emergence of a new unit of selection, the multicellular organism, from groups of cooperating cells. In this work, selection resulted from fitness interactions between cells within organisms and between organisms within populations. Cooperative interactions created fitness at the cell-group or emerging organism level. In addition, the model included parameters representing growth and development.

*Heritability.* Offspring must resemble parents for natural selection to occur. Although Darwin did not understand the mechanism of heritability, he did understand its critical importance to the process of natural selection. We saw in chapter 6 how heritability of fitness emerges at a new higher level out of the need to regulate conflict among lower-level units. Without this regulation there can be no transition to higher levels of complexity, because the heritability of fitness at the new level is decreased by the evolution of costly cooperative traits that lead to within-group conflict and the increase of defection. In studying individuality, I adopted the methods of quantitative genetics, viewing fitness as a quantitative trait, and defining heritability as the regression of offspring fitness on parent fitness. In other models, heritability at the appropriate level (for example, molecular replicator or organism) was represented by assuming that the offspring inherited their parents' so-called heritable capacities. In these models, a heritable capacity is a phenotypic trait tagged to the genotype that enters into a Darwinian dynamic as a type-specific constant. For example, the birth rate $b_i$ and death rate $d_i$ are measures of the heritable capacities of type $i$ at birth and death.[5] Such heritable phenotypic traits are a common, almost universal, feature of models in population biology.

Another way of expressing Darwin's insight is that for any system of reproducing entities in which there are heritable variations in fitness, natural selection must and will occur. In this statement of natural selection, "fitness" may be defined as an entity's expected reproductive success (an realized value), or in the sense of its capacity (by virtue of its traits) to survive and reproduce. In either case, "fitness" can be interpreted as representing Darwin's struggle to survive and reproduce.

Stating natural selection as a process resulting from certain conditions (heritable variation in fitness), with appropriate ceteris paribus clauses, rather than in terms of an expected outcome (for example, survival of the fittest) illustrates the lawlike stature of natural selection in biology. The claim is general and universal; it makes no reference to particulars like the planet earth; indeed, it applies to any system of replicators, whether they be on earth or on some other planet. The conditions apply to artificial replicators that exist in the memory of a computer as in the field of artificial life. The claim is not even restricted to living systems, and in this spirit my colleagues and I argued that Darwin's conditions apply to the emergence of order in certain physical systems that are far from thermodynamic equilibrium, such as lasers (Bernstein et al. 1983). Perhaps they apply also to categories and thoughts in the human brain and the nervous systems of other animals (Edelman 1987).

Endler (1986) concludes that natural selection should probably not be considered a law, since it primarily involves the application of the laws of probability. There is much more to natural selection and fitness arguments than the application of the laws of probability, as I hope I have shown in this book. The view that the varied uses and meanings of fitness can be boiled down to the use of probability is one of the central limitations of the so-called propensity interpretation of fitness discussed and criticized below. Furthermore, this view underestimates the empirical content contained in Darwin's conditions, as well as the very wide range of situations encompassed by them, from physical to living systems. I hope that the different examples of natural selection considered in this book lend credence to the view that natural selection is the central law and organizing principle in all of biology. Indeed, I think it is the only uniquely biological law.

Focusing on natural selection as a process contingent on certain conditions instead of considering it in terms of its expected products (such as better-designed organisms), allows us to handle apparent counterexamples in which Darwin's conditions appear to be met, but better designed individuals do not evolve. After all, laws should not be false, so how can we understand cases in which better-adapted types do not evolve by natural selection, that is, do not increase in frequency, even when Darwin's conditions appear to be met? It is well known that in small populations genetic drift can overpower the effects of selection and lead to less fit (poorly designed) individuals. As populations get small, more and more of the variation in fitness (reproductive and survival success) is not heritable. In other words, success depends more upon chance and less upon the heritable capacities of individuals. It is more difficult to satisfy Darwin's condition of heritability in small populations. Eigen (1971) referred to genetic drift as "survival of the survivors" by analogy with Darwin's "survival of the fittest."

The environment and the genetic and reproductive system can also reduce the heritability of fitness. The various constraints to adaptation stemming from the genetic system (such as heterozygote superiority, linkage, epistasis, or pleiotropy) can be understood in terms of their effects on heritablilty. Changes in the environment, whether due to random effects, the evolutionary responses of other co-evolving species, or local changes in fitness due to changes in frequency (when fitness is frequency-dependent) or behavior

171

(when behavior depends on learning or other conditions) all serve to reduce the heritability of fitness. That frequency-dependence reduces the heritability of fitness is especially obvious when there are negative frequency-dependent effect on fitness (or a cost of commonness) as discussed in chapter 4. When cooperation occurs, fitness (expected reproductive success) is maximum when all individuals cooperate.[6] However, natural selection cannot be expected even to approach this outcome unless additional conditions are met. In this book I have been especially interested in these conditions, as they promote the passage between evolutionary levels. Here I only wish to point out that variation in fitness need not result in an increase in fitness (either in the sense of better design or in the sense of expected reproductive success). In other words, natural selection is not synonymous with "survival of the fittest," in spite of Darwin's claim to the contrary. Indeed, once we understand the consequences of Darwin's conditions (in terms of the Darwinian dynamics that result) it comes as no surprise that the "fittest" types do not always increase under natural selection (recall the "survival of the common" and "survival of anybody" paradigms introduced in chapters 2 and 3 and discussed further below). For these reasons, maximizing fitness is an incomplete and potentially misleading approach to understanding natural selection.

Natural selection is a complex dynamical process with a variety of outcomes. Only confusion can result by associating it with a single outcome; the expected outcome most often being the maximization of fitness of the individual. The kind of confusion that can result is illustrated by Dawkins' argument for the gene as the sole unit of selection (Dawkins 1982), as I discussed in chapter 7 ("Reconsidering Fitness").

In the many attempts to analyze Darwin's theory of natural selection, two aspects have not been generally appreciated. First, in addition to natural selection in biological systems, Darwin's conditions also apply to the emergence of order in certain far-from-equilibrium nonliving systems (Bernstein et al. 1983). The fact that processes akin to natural selection occur in physical systems such as lasers that meet Darwin's conditions provides convincing support for the contention that natural selection is a law of universal generality. The second underappreciated fact about Darwin's conditions is that they provide for a rich spectrum of dynamical behavior (Eigen and Schuster 1979; Eigen 1971; Bernstein et al. 1984; Michod 1983b, 1995, 1991). These dynamics have led to a diversity of life forms, and to different kinds of individuals in the hierarchy of life (chapters 2, 3, 4, and 5). Furthermore, these dynamics may be used to classify natural selection into different selection paradigms, and to further clarify our understanding of the role of fitness in evolutionary dynamics.

172

## Paradigms for Natural Selection

Fitness often depends on population frequency and density. Density dependence of fitness is implicit in Darwin's struggle for existence. Resource regulation of population density is commonly represented in models by making resources a decreasing function of the density of the types in the population (see, for example, equation 2-2 or equation 3-2). However, this resource term considers only those density-dependent elements that are shared by both competing types. There are other within-species density-dependent components to the struggle for existence, such as the chances of encountering a protein (chapter 3) or the chances of mating in a sexual species discussed at the end of chapter 7. Elsewhere I have studied these nonlinear effects of within-species density dependence and how they affect the outcome of natural selection (see Michod 1991; 1995, chap. 9).

Birth and death are universal components of the life cycle. When birth and death are independent of density (and/or frequency), there is Malthusian (exponential) growth, and the outcome of natural selection may be described by the "survival of the fittest" or adaptationist paradigm, as discussed in chapters 2, 3, and 7. When birth and/or death are density dependent, there are four situations to consider at the outset, according to whether increased density increases or decreases the rate of birth or the rate of death. The mortality rate may decrease as density increases, if, for example, social interactions are important in protection. The mortality rate may increase as density increases, when type-specific disease is important, as when parasites or predators fixate on common types. This gives rise to a cost of commonness, or the "survival of anybody" paradigm, as discussed previously (Michod 1991; Szathmáry 1991) and considered in chapters 2 and 4. The birth rate may decrease when the species is rare, which is expected to occur as a result of sex or, as discussed in chapter 3, as a result of interactions with proteins in protein-mediated replication. This gives rise to a cost of rarity, or the "survival of the first" paradigm, discussed in chapter 3 and further explored at the end of the last chapter as a mechanism for creating species (Bernstein et al. 1985c; Hopf and Hopf 1985; Michod 1984, 1991). Finally, the birth rate may increase when the species is rare because of effects that are not included in the limiting resources, $R$. The rare-male mating advantage in *Drosophila* is an example (Anderson 1969). Also, under local mate competition, rare species may bias their sex ratio towards females, leading to an increase in reproductive rate when rare.

Selection is often classified according to the level at which it acts

(Lewontin 1970). Terms such as "group" or "individual selection" refer to this aspect. An equally significant feature of natural selection is the kind of dynamic involved. In the studies of natural selection given above, my emphasis has been on the dynamics of change, especially during evolutionary transitions. Classifying natural selection by the dynamics involved is not a substitute for the more common hierarchical approach, but provides new and significant information about the nature of natural selection. The dynamics of natural selection are as important as the units involved.

The dynamical approach distinguishes qualitative features of the evolutionary process associated with the different dynamics (for more discussion, see Michod 1991, 1995, chaps. 9 and 10). As already discussed in chapter 3 above, the dynamical approach was first used by Eigen (1971) and later by Eigen and Schuster (1979). Eigen and Schuster used the term "non-Darwinian evolution" to refer to the kind of dynamics resulting from a cost of rarity. As already discussed, I think this terminology is misleading, since there is nothing non-Darwinian about the dynamics studied in chapter 3; indeed, I have offered them as examples of Darwinian dynamics. Nevertheless, Eigen and Schuster contributed greatly to our understanding of natural selection by classifying it according to its dynamical character. My colleagues and I applied the dynamical approach to a variety of problems in evolutionary biology, including the origin of adaptation (Bernstein et al. 1983; see also chapters 2 and 3 above) and protein-mediated replication in simple systems of molecular replicators (Michod 1983b; see also chapter 4 above), the origin and evolution of sex (Bernstein et al. 1984; Michod 1995) and the origin and evolution of distinct species (Bernstein et al. 1985a,b; see also chapter 7, "Darwin's Dilemmas," above). The dynamical approach served as the basis for our study of the roles of cooperation and conflict in the emergence of individuality in chapters 5 and 6.

The three different patterns of natural selection—survival of the fittest, survival of the first or survival of the common (cost of rarity), and survival of anybody (cost of commonness)—can be understood as different outcomes of the Darwinian program. The fittest, first, and anybody paradigms are different outcomes of Darwinian dynamics that embody Darwin's conditions. Undoubtedly, there are other paradigms waiting to be discovered.

## FITNESS IN DARWINIAN DYNAMICS

To understand clearly Darwinian dynamics and the relation between the three cases just mentioned, it is necessary to understand the meaning of "fitness." As introduced in chapter 1, Fisher proposed measuring fitness by a genotype's rate of increase—the "objective fact of representation in future generations" (1958, p. 37). If difference equations are used in place of differential equations, Fisherian fitness becomes the finite difference in frequency or density of a type. Following Fisher's lead, Henry Byerly and I defined a set of fitness terms emphasizing different strands in Fisher's approach (Byerly and Michod 1991a). The actual rate of increase of a type includes both systematic factors like natural selection and random factors like genetic drift. Since we want a concept of fitness that applies only to the systematic process of natural selection, I restrict the discussion of fitness to $F$-fitness ($F$ after Fisher), defined as the per capita rate of increase as it is causally determined by a genotype's heritable capacities, reproductive system, and genetic system in systematic interaction with the environment. Most of the dynamical equations studied in this book are examples of $F$-fitness functions (see footnote 2 above).

There are, however, a variety of other fitness concepts in population biology, for example, Darwinian fitness, Wrightian fitness, genotypic fitness (for example, equation 4-17), selective value, and inclusive fitness (for example, equation 4-20), to name a few. In the modeling of natural selection, we seek a fitness concept that embodies the systematic components of the transformation in frequency of a type during a single generation.[7] The other fitness concepts just mentioned involve intervening steps in the determination of the rate of increase of a type during a generation. In populations with overlapping generations there are other fitness concepts (Charlesworth 1980), often with genotype-specific analogs, including the intrinsic rate of increase (defined by the Euler equation), net reproductive rate (expected reproductive success), and reproductive value (the conditional expectation of future reproductive success weighted by the growth rate of the population). Most central to the models used in population genetics and evolutionary ecology is the concept of individual fitness.

## THE INSUFFICIENCY OF INDIVIDUAL FITNESS

A common element of many evolutionary studies of natural selection is individual fitness. Indeed, this was Darwin's focus. Most philosophical discussions of natural selection concentrate on defining individual fitness and understanding its use in evolutionary explanation. By "individual" fitness we do not mean to refer to the fitness of a particular individual (which is, of course, of little concern to the dynamics of natural selection in a popula-

Shortcomings of Individual Fitness

Not dynamically sufficient
Not maximized or increased by natural selection
Statistical construct, need not apply to any particular individual
Lumps effect at different levels
Placeholder for causal analysis
Differences in individual fitness are not necessary or sufficient for
　　natural selection

*Table 8-1*

Shortcomings of individual fitness in evolutionary explanation. Individual fitness is defined as expected reproductive success of a genotype class. See text for explanation.

tion), but rather to the fitness of a class of individuals with the same trait. In genetical models class is designated by genotype, say $i$, with individual fitness $W_i$ defined as the expected number of gametes produced by individuals with genotype $i$ ("expected" means taking into account survivorship). As considered in chapter 4, most population genetic models use genotypic fitness in calculating the rate of increase of a gene or genotype (as in the case of the simplest life history model with discrete, nonoverlapping generations).

Many examples of individual genotypic fitness have been considered in this book. Recall the $W$ terms in chapter 4, or individual fitness defined in table 5-2 and used throughout our study of individuality in chapters 5 and 6. Individual fitness can be a useful intervening fitness concept because it summarizes certain critical components of the life cycle of the individual (survival and fecundity), the individual being defined by its genotypic class.

Although a useful construct, individual fitness has significant shortcomings in evolutionary explanations and predictions based on natural selection (table 8-1). Most critically, individual fitness is dynamically insufficient, as it does not contain all the components involved in natural selection, even when considering change during a single generation, especially when there is sexual recombination, or when there is frequency-dependent selection in a multilevel setting. Individual organisms are often considered as the optimum, or fitness-maximizing, units in evolution. However, we know that individual fitness may decrease as a result of frequency-dependent selection when there are interactions. For example, individual fitness ends up minimized if it depends upon interactions akin to the prisoner's dilemma game, if interactions only occur once (chapter 4, "Prisoner's Dilemma"). Another problem is that individual fitness is sometimes a statistical construct. For example, in the case of kin selection, no individual organism actually has the individual fitness measured by, for example, equation 7-1.

Sometimes individual fitness lumps together components of change that are best kept distinct. For example, in the multilevel setting used in chapters 5 and 6, individual fitness combined between- and within-organism change. It is the increased covariance of individual fitness at the new emerging level with genotype that drives the evolutionary transitions studied in chapter 6 (recall figure 6-4). For this reason, in chapter 6, I focused on the heritability of individual fitness as the defining characteristic of an evolutionary individual, since heritability of fitness captures the aspects of selection most relevant to the new emerging unit. For similar reasons, the Price equation 5-2 partitions the total change occurring in a single generation into a heritable covariance component and within-level change.

Maynard Smith (1991b) criticized a previous paper of ours (Byerly and Michod 1991a,b) because we used $F$-fitness instead of the more commonly used individual (genotypic) fitness in our analysis of the role of fitness in evolutionary explanation.[8] Our goal in that paper (like mine here) was to develop a schema for fitness explanations that reflected the use of models in evolutionary explanation in an epistemologically correct way. We used Fisher's $F$-fitness instead of individual fitness because we wanted a schema that represents the whole process of natural selection, one that embodies all of Darwin's conditions of variation, heritability, and the struggle to survive and reproduce. Although individual fitness summarizes the survival and fecundity (gamete production) components of the life cycle, it falls short of predicting the course of evolution under natural selection, because it leaves out genetic transmission and the heritability component, which is one of Darwin's central conditions. Furthermore, individual fitness itself is often just a placeholder for more explicit models of causality based on individual heritable capacities and the environment, as I discuss in more detail below when I consider the propensity interpretation.

Genetic transmission is always relevant, but it is especially relevant for the kinds of evolutionary problems considered here, such as when genetic relatives are present in the population (see chapter 4, "Kin Selection," and chapter 7, "Kin Selection of Altruism"), when there are interactions at multiple levels (as in the model of multicellular organisms in chapters 5 and 6 or in the evolution of social behavior), when there is nonadditivity of allelic effects (as when there is heterozygote superiority; see chapter 7, "Heterozygote Superiority"), when there are transitions in reproductive mode (chapter 7, "Cost of Sex"), or when reproductive success depends upon the mating pair.

When selection is occurring on several levels simultaneously, as is often the case, fitness at any one level must often be a poor guide to understanding the outcome, even if the level is the organism. Recall in chapter 5 how the individual fitness of the cell group (the organism) was quite sensitive to changes in the benefit of cooperation, β (panel (C) of figure 5-4), as was the initial increase of cooperation genes (figure 5-2). Nevertheless, the overall outcome of selection, in terms of the level of cooperation ultimately attained in the organism, was relatively insensitive to the effect of cooperation on individual fitness, β (panel (C) of figure 5-3).

Average individual fitness is equally limited in explaining evolution. The main point of Fisher's (1941) treatment of selfing was that intense selection need have no effect on the well-being of the population or individuals within the population (chapter 7, "Evolution of Selfing"). Average individual fitness is also of limited use in understanding frequency-dependent natural selection (recall equations 4-5, 4-8, and 4-24), and was insufficient to understand the evolution of cooperation among cells in the transition from single cells to multicellular organisms (recall figures 5-3 and 5-4).

Differences in individual (genotypic) fitness are neither necessary nor sufficient for interesting evolution by natural selection. The simplest counterexample to the claim of sufficiency is the case of constant selection with heterozygote superiority in individual fitness. In this case differences in individual fitness always exist, but there may or may not be evolution, depending on whether the population is at equilibrium. Furthermore, at no point in the evolution of a population does the fittest genotype (by definition the heterozygote) have the highest per capita rate of increase—the fittest do not survive the best over evolutionary time (see the last three sections of chapter 7, or Michod 1995, chap. 9 for more discussion).

The reason that "survival of the fittest" is false is easy to understand in the case of heterozygote superiority: the fittest genotype, the heterozygote, cannot breed true because of sex. Because of mating and recombination, the heterozygote always segregates the less fit homozygotes among its offspring. When reproduction is sexual, an organism's offspring are a mixture of different genotypes and, consequently, a genotype's output includes a contribution of gametes (or offspring) to other genotype classes. In the example of heterozygote superiority discussed in chapter 7, the genotypic fitnesses are constant by definition. Nevertheless, the population goes through phases of rapid evolution and stasis when it is at equilibrium. At equilibrium, even though there are differences in the survival and reproduction of organisms, other components of the process of natural selection (mating and recombination) are working to keep the population in check.

Differences in individual survival and reproduction are not sufficient for there to be natural selection. Are they necessary? It might seem that if natural selection occurs there must have been differences in individual fitness. However, this too is not true. Recall the intrinsic advantage of parthenogenesis over sexual reproduction. In this case, it is assumed that the survival and number of gametes produced is the same for all types, yet there is powerful natural selection because of effects of the mating system. The effects of the mating system are not properties of the individual genotypes alone but involve density- and frequency-dependent opportunities for successful mating and propagation of the underlying genetic elements, in other words, differences in the heritability of fitness.

## HERITABILITY AND NATURAL SELECTION

Natural selection is more than differential survival and reproduction of individuals, it requires these differences to be heritable. According to Price's general covariance approach introduced in chapter 1 and used throughout chapters 5 and 6, natural selection includes both the component of assortment of classes and the component of property change, and both of these components are affected by sexual reproduction and within-level change (Price 1995). To predict the outcome of natural selection during the transition from cells to multicellular organisms, individual fitness was weighted by heritability.[9] Likewise, the two components of the Price equation (e.g., equation 5-2) (the covariance term which represents the mapping of heritable components in the zygote into adult fitness and the second term which includes the within-organism change) partition the total change in a multilevel selection hierarchy. Heritability of fitness provides the tools needed to understand the emergence of cooperation and the individuality of organisms (see footnote 9). In chapter 5 we found that greater levels of cooperation and harmony for the organism may be attained when the heritable covariance of fitness at the organism level overpowers the within-organism change toward defection.

Recall in chapter 5 how, as the within-organism selection parameters ($b$ and $t$) increased, individual fitness declined (figure 5-4). Nevertheless, the level of cooperation in the organisms and population was relatively unaffected by changes in $b$ and $t$ up to the limiting values (figure 5-3). We found that the deleterious effects of within-organism change on individual fitness

179

were compensated for by an increased heritability of fitness at the organism level (figure 5-5)—so that the existing level of cooperation and harmony within the organism was preserved (figure 5-3), even though the average organism fitness declined (figure 5-4, panels (A) and (D)).

Heritability of fitness provides the necessary concept to fully understand the counterexamples to using individual fitness for predicting the outcomes of natural selection. Finite heritability of fitness is required for evolution by natural selection. Heritability of fitness drops to zero at equilibrium since, although there remain differences in survival and reproduction of individuals and genotypes, these differences are no longer heritable at equilibrium (as in the case of heterozygote superiority). Consequently, natural selection is not occurring when the population is at equilibrium, even though some individuals (the heterozygotes) are surviving and reproducing better than others. In our consideration of the cost of sex, although the fitness of sexually and asexually produced offspring is assumed to be the same, the heritability of fitness is lower in sexual populations by a factor of approximately two.

There are two related ways to understand the tension between individual fitness and evolutionary success in these examples (heterozygote superiority and the evolution of reproductive mode). First, for the selected type (asexuals, selfers, homozygotes) the heritability of fitness is higher. Second, we might say that selection is occurring at a lower level than the individual organism, at the level of the gene. Both views seem correct to me, although to focus on the lower level can be misleading in regards to understanding the *causes* of the selective differences. Genes don't exist alone, they gave up that carefree existence some time ago; now they come packaged in other units such as organisms with their reproductive and other traits. It is the genetic and reproductive system of the organism that enters into the Darwinian dynamic and is reflected in terms of greater heritability of the selected type.

Some philosophers have argued for separating individual fitness (expected reproductive success) and heritability in the process of natural selection (Brandon 1991; see below).[10] If the break between selection and heritability were clean, I would go along with this view. As Brandon argues it would enable us to take a two-step approach to evolutionary change, the first step being the selection of phenotypes and the second step the transmission of phenotypes across the generations. Quantitative genetics seems to recommend this approach in its separation of selection from the response to selection. The examples studied in this book teach the opposite, that genetic transmission and phenotype survival and reproduction are inextricably intertwined, especially in situations in which selection is occurring simultaneously on several different levels. Multilevel selection must be occurring when there are transitions in units of selection, or when there are effects of parasites and other lower-level conflicts on the reproduction of whole organisms (see "Reconsidering Fitness" in chapter 7).

The view that "selection" (the act of selecting) can be separated out from the process of natural selection (involving heritability) stems, in part, from the (misleading) analogy between natural selection and artificial selection. Artificial selection shares with subset selection (everyday selection in which someone chooses, that is, "selects," an item from a set) a fundamental property that natural selection does *not* have: there is a selecting agent and so the selection step is distinct. Recall Price's general schema for selection discussed in chapter 1 (Price 1995). In the case of natural selection, the offspring present in the next generation result from both differential survival and reproduction and, when there is sex or mutation, property change (the offspring may have different properties from the parents). The two components of natural selection do not always occur in sequence, but are often inextricably intertwined, especially when there is multilevel selection or selection on reproductive traits. Evolutionary biologists often speak as if the environment "selects." In some cases, as with Kettlewell's moths in industrial England, the environmental challenge seems clear and sex or levels of selection conflicts don't distort the signal, but in other cases, as when there are genomic conflicts, the evolution of reproductive mode, or transitions between levels of selection, the analogy breaks down.

We cannot generally expect the effects of "selection" to be distinct from the contributions of heritability. Any attempt at a general formulation of natural selection that begins by partitioning phenotypic selection from heritability seems fated from the beginning to be of limited generality. This approach may work if the units of selection are well defined, but in cases where the units of selection are being formed, when there is manipulation, or when there is a breakdown in the integrity of a level of selection (and there are more examples of these processes every day), selection and heritability are inextricably intertwined. This is especially true in the evolution of parental effects on the expression of traits in the offspring, as occurs for a wide variety of traits.

## SCHEMA FOR NATURAL SELECTION

An explicit formulation of the full dynamics of natural selection in nature is probably impossible. There are many kinds of approximations necessary

181

in models and in explanations. Nevertheless, it is possible to provide a general schema for natural selection. Consider, for starters, the case of Malthusian growth in the context of the following familiar equation from population ecology[11]:

$$\frac{1}{X_i} \frac{dX_i}{dt} = b_i R - d_i. \tag{8-1}$$

Equation 8-1 exhibits a simple form of the functional dependence of $F$-fitness on heritable capacities and environmental variables. It is identical to Fisher's deconstruction in equation 1-2 except for the resource term.

It is important to distinguish the $F$-fitness value itself from the factors that causally determine it. The $F$-fitness value, for example, the left-hand side of equation 8-1, is a real number giving the per capita rate of increase of genotype $i$. The factors that causally determine this value are given on the right-hand side of the equation. The right-hand side exhibits the functional dependence of $F$-fitness on the heritable capacities: birth rate $b_i$, death rate $d_i$, and $R$, the resources available in the environment. The capacity for birth, $b_i$, can be operationally defined as the birth rate at abundant resource levels. The realized birth rate, $b_i R$, takes into account the limiting effect of available resources.

Changes in relative frequency of genotypes correspond to evolutionary changes within a population. Within the model represented by equation 8-1, given measures of the heritable capacities $b_i$, and $d_i$ and resources $R$, one could predict the outcome of selection by iterating equation 8-1. However, in this case, the ratio $b_i/d_i$ predicts the outcome of selection (as suggested by Fisher's deconstruction given in equation 1-2). Since the outcome of selection depends only on the heritable capacities of each type, within this model one could, in principle, predict the evolutionary success of a genotype's phenotype from its "design" as measured by the ratio of the birth and death capacities. We then have evolutionary success of the better designed, that is, survival of the fittest. This is the basic feature of the survival of the fittest dynamic: adaptiveness at specified activities predicts evolutionary success. In the many counterexamples to this reasoning given in this book, adaptiveness as represented by some combination of heritable capacities is not sufficient to predict the outcome of selection, but evolutionary change is still determinable, within limits of chance factors, by the $F$-fitness function itself.

The dynamics of equation 8-1 represent the simplest possible case of selection. Replication is asexual, the phenotype essentially coincides with the genotype, and the dominant interaction with the environment involves re-source utilization. Sexual reproduction complicates the dynamics, but Darwin's conditions still hold. In more specific models, other heritable capacities are studied, such as the fitness effects traded during the evolution of interactions. For certain problems, the birth and death rate terms, $b_i$ and $d_i$, may require further analysis in terms of interactions among other heritable capacities (involving, for example, the mating, social, or sexual system) and environmental factors (such as density and frequency of genotypes). Such variations on the basic theme inherent in the Darwinian dynamic, point, I think, to a virtue rather than a defect in its conceptualization.

A general schema for the dependence of $F$-fitness on heritable properties of individuals $a_i$ and environmental factors $E$ (using vector notation to emphasize multiple capacities and environmental variables and parameters) is given in equation 8-2, in which the function $F_i$ takes into account the genetic and reproductive system (Byerly and Michod 1991a):

$$\frac{1}{X_i} \frac{dX_i}{dt} = F_i(\mathbf{a}_i, \mathbf{E}).$$  (8-2)

The environmental factors, **E,** may include the frequencies, densities, and heritable capacities of other genotypes in the population. Complex selection processes at different levels can be schematized in terms of relations among the same kinds of factors, as we have done in chapters 4 and 5. These factors are the densities of classes of individuals in a population, $X_i$, heritable capacities of individuals that make up the classes, $\mathbf{a}_i$, and environmental factors, **E.** These factors are combined in the function $F$ in a way dictated by the hypothesized interactions and the assumed genetic and reproductive systems.

## The Propensity Interpretation of Fitness

A widely espoused attempt in the philosophical literature to understand fitness and natural selection is the so-called propensity interpretation of fitness. The fitness propensity of an individual $X$ in an environment $E$ is defined as the "expected number of descendants which $X$ will leave in $E$" (Mills and Beatty 1979). The propensity interpretation was developed initially to directly answer the tautology challenge discussed in chapter 7. The propensity interpretation views fitness as a probability and focuses on philosophical issues involved in the interpretation of probability. According to this view, random factors are the only source of nontautological definitions of fitness. Even though a type has the highest fitness propensity, chance factors may intervene to keep it from succeeding in survival and reproduction. Consider two identical twins, one being hit by lightning as a youth and the other living to old age with many children. We say that the fitness propensity of each is the same, even though they have different numbers of children.

The distinction in the propensity interpretation between fitness as a probability of reproductive success and actual reproductive success is well taken, but the analysis is inadequate as a theory of fitness in several respects (table 8-2). First, the varied use and meaning of fitness in evolutionary biology cannot be understood simply by understanding the various meanings of probability in philosophy. Second, the propensity interpretation obscures the causal connections that biologists actually make (between heritable capacities, genetic and mating systems, and the environment) by combining dispositions and capacities that should be kept distinct into one grand capacity, "fitness." Third, and this is related to the other shortcomings, it provides an incomplete answer to the tautology challenge. Finally, in the propensity view, fitness is really a placeholder for more complete descriptions of natural selection.

While I agree that chance events decouple actual evolutionary success from the adaptiveness of individual traits (even the best-designed organism may get struck by lightning), the propensity interpretation is misdirected in its focus on chance events as the primary source of genuine nontautological evolutionary explanations. If there were no chance factors affecting actual evolutionary change, the explanatory power of natural selection would only be strengthened. However, the propensity interpretation implies just the opposite. Are we really to believe that fitness explanations would be

---

Shortcomings of Propensity Interpretation of Fitness

Lumps capacities and propensities that are best kept distinct
Views fitness solely as a probability
Focuses on chance as the sole source of nontautological definitions
    of fitness
Placeholder for causal explantion

*Table 8-2*

Shortcomings of propensity interpretation of fitness. See text for explanation.

truly tautological in a deterministic world? A central point of the many examples of natural selection considered in this book is that fitness effects at multiple levels, genetic factors such as dominance and pleiotropy, ecological factors such as density and frequency of types, and reproductive factors such as sex, all intervene between individual capacities and evolutionary success. Chance intervenes too, but it is only one of many factors that may decouple individual capacities from evolutionary success.

An analogy that highlights the way stochastic factors enter into the selection process is a sequence of tosses of a die. The actual observed frequency in a sequence of tosses of the dice is analogous to the actual rate of increase of genotypes including random factors. The a priori probability calculated using physical principles and properties of an individual die correspond to the functional (systematic) dependence of $F$-fitness on heritable capacities in interaction with environments. A bias in a die (say, its shape, or being magnetized) is an analog of the heritable capacities of an individual. The determination of which side of the die turns up by relations between the properties of the die and conditions of the tossing (the environment) corresponds to determination of $F$-fitness.

Mills and Beatty (1979, p. 43) and Sober (1993, chap. 3) argue that the individual fitness propensity ascribed to an organism explains its individual fitness success (expected reproductive success) just as the property of solubility of salt in water explains its actually dissolving in water. We could equally well ascribe an $F$-fitness propensity to the genotype, but it is not useful to construe the value as reflecting a dispositional property of the organism. $F$-fitness values can be viewed as supervening on propensities of various traits of individual organisms (there are an infinite number of $F$-fitness functions that give the same $F$-fitness value). But we cannot add such propensities together to get an overall propensity, as it were, of an individual organism for evolutionary success. No further overall propensity or disposition intervenes between the heritable capacities and the rate of increase.

I don't see how it helps to observe that an organism has high fitness (in the sense of success) because it has a high propensity for fitness. It is a lot like the character in Molière's play *The Imaginary Invalid* who claims to explain why opium puts people to sleep by saying opium possesses a dormitive virtue. I have yet to read a satisfactory defense of the propensity interpretation of fitness that addresses this issue. For example, Sober, in an otherwise lucid account of fitness, embraces the propensity interpretation, even acknowledging that "To say that an organism's fitness is its propensity to survive and be reproductively successful is true but rather unilluminating" (1993, p. 66).

185

It is only by delving deeper into the biological and physical causes of this propensity that we are able to explain the evolutionary success of a genotype—and this is precisely what the $F$-fitness function does. The $F$-fitness function explicitly expresses the functional hypothesis about the traits under study, in the context of the genetic and reproductive systems. Sober seems in agreement with this goal, for he goes on to say (p. 74), "Once this physical characterization is obtained, we no longer need to use the word 'fitness' [read fitness as a propensity throughout this quotation] to explain why the traits changed frequency. The fitness concept provides our initial explanation, but the physical details provide a deeper one. This does not mean that the first account was entirely unexplanatory. Fitness is not the empty idea of a dormitive virtue. The point is that although fitness is explanatory, it seems to be a placeholder for a deeper account that dispenses with the concept of fitness [again the propensity concept]." I could not agree more with what Sober says here about the need for going deeper than fitness as a propensity, but having acknowledged this, I do not see in what sense he can regard the propensity concept as explanatory. This "initial explanation" is no explanation at all. As Sober says, it is simply a placeholder in need of more careful investigation.

I think this is what Byerly and I mean to accomplish when we bypass individual fitness in our discussions of natural selection, and focus directly on the per capita rate of increase, $F$-fitness, as a measure of evolutionary success expressed as a function of heritable capacities and aspects of the environment and the genetic and reproductive systems (Byerly and Michod 1991a). Individual fitness is a placeholder, an intervening construct, and must be deconstructed if we are to understand its causal basis in individual properties and its effects in terms of natural selection. The $F$-fitness *function* (the right-hand side) provides a sufficient causal model and hypothesis for evolutionary success. Once the $F$-fitness function is specified, there is no longer any need for a placeholder concept of fitness.

## Brandon's Approach

Brandon offers a treatment of fitness based on the propensity interpretation, and in the process criticizes a previous discussion of fitness that served as a basis for the framework presented here (see especially Brandon 1991, chap. 1). He finds my approach unnecessarily complicated and potentially misleading; unnecessarily complicated because it lumps differential survival and reproduction of phenotypes with genetic transmission; potentially misleading because it posits genotype-specific capacities that actually depend upon the environment.

Let me begin by acknowledging a point of agreement. Brandon notes that, in equations like equation 8-2, I should write $F_i(\mathbf{a}_i, \mathbf{E}, G)$ to emphasize that the per capita rate of increase is a function of the genetic system $G$ in addition to the individual capacities $\mathbf{a}_i$ and environment $\mathbf{E}$. He is certainly correct that the genetic system and heritability enter into $F$-fitness. My goal, after all, is to include all of Darwin's conditions into a schema for natural selection. Byerly and I understood the genetic system to be included in the $F_i$ function, because it affects the way the heritable capacities are combined (Byerly and Michod 1991a,b). In any event Brandon and I both agree that $F$-fitness depends upon the genetic system.

A substantial difference between us is that Brandon wishes to separate the notion of fitness from considerations of the genetic system and transmission. Let me say he has good company. Recall that most models (including those studied here) and empirical investigations use individual genotypic fitness, and Fisher himself sought a similar decomposition of natural selection in his fundamental theorem (see chapter 4). The question at issue is not just whether the notion of fitness should stop with the selecting step or should be extended to include the cross-generation effects involved with the transmission of phenotypes, but whether there even exists a clearly distinguished selecting step. When the phenotypes involve reproductive mode, parental effects, or interactions in structured populations, when there are transitions in units of selection, I think the commonsense notion of selection as involving a choice or subset according to certain criteria is a poor description of the actual dynamics involved, because of modifications of the properties of the entities themselves during transgenerational change. In these cases, the concept of fitness must be extended into the realm of transmission and heritability. The challenge is how to do it.

In our considerations of natural selection, we have combined heritable capacities of the organism, the genetic and mating systems, and the environment in the dynamics of evolutionary change. Brandon argues to separate individual survival and reproduction from the transmission of phenotypes in a two-step process, "selection" and the "response to selection," as suggested by the approach of quantitative genetics. By analogy with artificial selection, he uses "selection" to refer to the "selection itself," meaning the differential survival and reproduction of phenotypes, and by the "response to selection" he means the cross-generation genetic transmission of the phenotypes. I have already discussed this matter in chapter 1, "Selection as Fitness Covariance," and in the present chapter above, "Heritability and Natural Selection." If there were a clean break between these two processes, then it would be simpler to separate them. However, I hope the examples considered in this book make clear that in most interesting cases the two processes can be inextricably intertwined, especially during the evolution of reproductive mode or the emergence of new levels of selection. Furthermore, as discussed in the section "The Insufficiency of Individual Fitness" above, differences in survival and reproduction among organisms are neither necessary nor sufficient for interesting evolution to occur. How are we to understand the evolution of asexual reproduction (cost of sex), gynodioecy, selfing, or the emergence of new levels of selection (and hence new levels of fitness) according to this view?

In the case of selfing and asexual reproduction, we assumed there were no differences between phenotypes in their survival or number of gametes produced, yet we discovered powerful selection nevertheless, according to the differing opportunities for successful mating and propagation of the underlying genetic elements under the different mating systems. In the case of heterozygote superiority, there are differences in survival and reproduction, yet no evolution occurs when the population is at equilibrium. The propensity of the sterile worker bee to have offspring must be zero, yet we know its phenotype can evolve under natural selection if kin are present. In all cases of multilevel selection, including the passage from one level of selection to another, the fitness of the higher-level units must include many generations of the life cycle of the lower-level units. In addition, fitness effects are traded from the lower level to the higher level during the emergence of fitness at the new level. In other cases of multilevel selection involving parasites, the reproductive behavior of the host organism may be modified by the parasites, and fitness is traded in the reverse direction from the higher level to the lower.

Brandon's second criticism is that the genotype-specific parameters (indexed by letters $i$ and $j$ in the tables above) must in practice depend upon the environment.[12] Now, in an absolute sense he has to be right, everything is affected by everything else to some degree; but in a relevant sense this is a red herring—some things are more affected than others. If there were no heritable capacities, there could not be evolution by natural selection. So we know that heritable capacities exist. Recall, after Darwin, that heritability of fitness is a requirement for natural selection; it must be possible to assign heritable capacities to individual units of selection. The problem is to do it correctly and in the most heuristic way.

Another concern of Brandon's is environment-by-genotype interaction, which is a simple matter to handle in the general scheme proposed above.

Even in the case of equation 8-1, significant interactions of the genotype with the environment would be handled by making the birth parameter $b_i$ a function $b_i(\mathbf{a}_i, \mathbf{E})$ of relevant environmental variables and other heritable capacities. This is precisely what is done in models of sexual selection (see, for one of many examples, Michod and Hasson 1990).

## HERITABLE CAPACITIES AS COMPONENTS OF DESIGN

In models in population biology, heritable capacities represent the design attributes of the individual that interact (to some degree independently) with the environment and reproductive and genetic systems to determine evolutionary success. Heritable capacities may represent specific traits, such as the cost of making a protein (studied in chapters 2 and 3) or the length of a male's tail in a model of sexual selection. Heritable capacities may also be more complicated measures of individual attributes and behaviors, as with the costs and benefits of cooperation studied in chapters 5 and 6. In all cases, heritable capacities measure and represent design attributes of the individual which, when combined with the environment and the genetic and reproductive systems, determine $F$-fitness, that is, evolutionary success.

In philosophical discussions, fitness is often said to supervene on physical properties (Rosenberg 1985, 1978; Sober 1993). Supervenience refers to a many-to-one asymmetrical relation between properties of an object (in our case an individual) in which "One set of properties $P$ supervenes on another set of properties $Q$ precisely when the $Q$ properties of an object determine what its $P$ properties are—but not conversely" (Sober 1993). In the case of fitness, philosophers say that fitness supervenes on the biological and physical properties of individuals, because there are many biological and physical properties that may result in the same fitness. In Sober's words, "A cockroach and a zebra differ in numerous ways, but both may happen to have a 0.83 probability of surviving to adulthood."

Heritable capacities supervene on biological and physical properties. For example, in equation 8-1 for Malthusian growth, the birth capacity, $b$, is defined at the birth rate at unlimiting resources. Although this birth capacity is measurable in most cases, there are many underlying traits that determine this capacity (and these traits are likely to be different for different kinds of individuals). This is also true if we push the causal (functional) analysis further and express the birth capacity, $b$, as a function of other traits and capacities, such as tail length in male peacocks, if we were interested in a problem involving attraction between mates and sexual selection.[13] By ex-

189

pressing the birth capacity as a function of tail length (and perhaps female choice), we are being more explicit in our explanation of natural selection; nevertheless, tail length supervenes on other physical and biological properties, because there are many ways of getting a particular tail length. No matter where we stand in the causal nexus we could go further. Where we stop depends upon the nature of the question being asked.

A dividing issue between the account of fitness offered here and the propensity account discussed in the previous section is whether capacities of the individual can be combined into an overall capacity or propensity (call it overall adaptedness) which determines evolutionary success ($F$-fitness). Evolutionary success includes across-generation change; however, the same issues exist with regard to success at survival and reproduction within a single generation. The propensity interpretation of fitness claims that it is generally possible to combine individual attributes into an overall propensity that explains evolutionary success (or success at survival and reproduction). However, for most of the cases studied here it has not been possible to combine the heritable capacities into an overall individual capacity or propensity that predicts success. The propensity account says that an individual's overall propensity for reproductive (or evolutionary) success explains its reproductive (or evolutionary) success. The $F$-fitness account given here says that an individual's heritable capacities interact with the environment in the context of the genetic and reproductive systems to explain evolutionary (or reproductive) success.

Although both the propensity account and the $F$-fitness account refer to individual capacities and propensities, I believe there are important differences between these accounts of the role of fitness in evolutionary explanation. An underlying principle that emerges from the $F$-fitness account is that an individual's heritable capacities cannot typically be coalesced into one grand capacity or propensity. This principle embodies the deconstruction of fitness so crucial to modern evolutionary and ecological thought. Evolutionary individuals are not unanalyzable wholes, nor are they a simple sum (or some other function) of their parts, that is, their traits and capacities. Individuals are to a significant extent decomposable and modular; indeed, decomposability of capacities and their underlying traits is necessary for continued evolution (otherwise the slightest change in one trait would destroy the functioning of the whole).

Different individual capacities interact with the environment, to some degree independently, as distinct causal components of $F$-fitness, and once these interactions occur there is usually no way of returning to an overall

190

property or propensity of the individual that predicts the outcome of selection. Are we really to believe, as the propensity interpretation implies, that it is "fitness" (viewed as an overall propensity of individuals) that interacts with the environment to yield reproductive success? I think not. Instead, we should see fitness as the *result* of interactions of the individual's heritable traits with the environment in the context of the genetic and reproductive system.

Fitness is an effect of interactions, not a cause. During a single generation, fitness is constructed out of individual capacities in interaction with the environment (in the context of the genetic and reproductive systems). During evolutionary transitions, new levels of fitness are constructed out of interactions among lower-level units and other population processes, as I have taken care to explain in the preceding chapters.

The $F$-fitness *function* combines an individual's heritable capacities and the environment in the context of the genetic and reproductive systems into an overall statement or hypothesis of evolution by natural selection. One could imagine a global propensity for reproductive or evolutionary success, but there is little point in doing so when one has an adequate model of natural selection, that is, an $F$-fitness function for the problem at hand. There is a real difference in causal adequacy between combining heritable capacities of individuals in an $F$-fitness function of the environment and genetic and reproductive systems as opposed to positing a single overall fitness propensity. The latter is just Molière's dormitive virtue, while the former is a scientific hypothesis about the causal components underlying the dynamics of natural selection.

## Overall Adaptedness of Organisms

In spite of the problems inherent in viewing fitness as an overall property of the organism, fitness is often interpreted in this way, as overall adaptedness, by biologists and philosophers alike, and the propensity interpretation follows in this well-established tradition (see table B-1, quotations 10, 11, 12, 16, and 17). The conception of fitness as overall adaptedness emphasizes the design of an organism to survive in its environment rather than its actual survival and reproductive success. But without an underlying model of how the design works, in interaction with the environment and the genetic and reproductive systems, there can be no explanation. The $F$-fitness function (in other words, the right-hand side of equation 8-2 or any of the

191

other many examples considered in this book[14]) provides such a mechanistic model of design in the context of the question at hand and the environment and the genetic and reproductive systems.

The misleading suggestion in using "adaptedness" for "fitness" becomes obvious if we ask: Is the adaptedness of a sterile worker bee zero because it has no offspring? And if so, how can we reconcile this conclusion with the fact that natural selection may favor such behavior if the inclusive fitness conditions are met? The sterile worker bee is well adapted at delivering the benefits posited in the kin selection model. Even though its individual fitness is zero, its behavior and the underlying genes may increase (see equations 4-20 and 4-21).

Construing "fitness" as overall adaptedness of an organism leads Rosenberg (1978, p. 369) and Sober (1984, p. 74) to puzzle unnecessarily over the question: How can fitness explain differences in rates of increase if fitness values are measured by rates of increase? The simple answer is: $F$-fitness values do not *explain* rates of increase, they just *are* such rates. It is the $F$-fitness function (the right-hand side) that explains how the $F$-fitness values (the left-hand side) are causally determined (see note 14).

I discussed the insufficiency of individual fitness above. The problems discussed there highlight the role of heritability and the complexity of natural selection when fitness is frequency dependent. How do these issues bear on whether individual fitness can be used as a measure of overall adaptedness of the organism, as the propensity interpretation of fitness proposes? There are two basic problems with viewing individual fitness as a measure of overall adaptedness of the organism (also see table 8-1). First, individual genotypic fitness, like $F$-fitness, is not usually a property of the individual alone, or even the genotype, but a property of the genotype-in-a-population. Secondly, in many cases, for example in models of kin selection (equation 7-1), reciprocal altruism (see, for example, Brown et al. 1982), or in game theory models, individual fitness is a statistical construct used in the process of calculating the outcome of natural selection but is not a property of any particular organism in the population.

There is no single property that serves as a common physical basis for the disposition posited by the propensity interpretation, the propensity for a type of organism to have an $F$-fitness value (or individual fitness value) in a given environment. There is no simple set of regularities for summing heritable capacities into an overall adaptedness that predicts the outcome of natural selection, by analogy with the parallelogram law in physics for combining force or acceleration vectors into a resultant vector. There is, of course, the $F$-fitness function (the right-hand side) and the resulting value

192

(the left-hand side of equations like equation 8-1), but this is not a dispositional property of the individual. One might think we could abstract from the dependency of $F$-fitness on the heritable capacities $\mathbf{a}_i$ a "property" (call it $A_i$) which is a separate function of only the components of $\mathbf{a}_i$ that would be a candidate for overall adaptedness. This was possible for the case of template replicators replicating according to equation 2-2 or equation 8-1 (recall the selection conditions, equation 2-5); however, it is simply not possible to do this in general.[15] Yet, when reduced to its essence, this is precisely what the propensity interpretation claims is generally possible.

Overall adaptedness is not well defined even as a property purportedly supervenient on heritable capacities or phenotypic traits of organisms, since there is no common measure for such a concept. This is in effect what Fisher (1930, p. 39) notes in his remark that "fitness is qualitatively different for each different organism whereas entropy, like temperature, is taken to have the same meaning for all physical systems." The basis, the heritable capacities that make up $F$-fitness, is qualitatively different for every case study in natural selection.

To better appreciate the basic problem in trying to measure overall adaptedness, consider the analogy of adaptedness to good design of an automobile. We might observe that an automobile is well designed, say, for top speed, fast acceleration, durability, for comfort, or various other functions. And we can refer to specific properties, analogous to heritable capacities, that "adapt" the automobile to these functions. An engineer might construe the time taken to reach top speed as an $F$-fitness–like function of an auto's capacities (its horsepower, gear ratios, etc.). We can also put the automobile on the track and measure its acceleration and compare it to the value predicted by the $F$-fitness function as a means to check our understanding of the factors involved. This is what can be done; this is all that can be done, and it is quite enough. The various capacities of the automobile do not "add up" to any unified, overall "good design" property. To speak of the automobile having a "propensity" to have a given acceleration may sound good, but it doesn't add to our understanding of the factors involved.

That no collection of physical properties can be cited as a general basis for fitness does not, however, imply that causal explanations cannot be given. Indeed, all of the evolutionary dynamics studied here (see note 14), or the other numerous examples in population biology, are nothing if they are not causal explanations of the evolutionary process. $F$-fitness *values* supervene on the various causal relations in the selection process (the $F$-fitness *function*) in the sense that there is a common measure which is determined by a variety of biological interactions. The concept of $F$-fitness is

193

unified as an effect even though its values have a great diversity of underlying causes.

## MASKING OF ADAPTIVENESS

Predictions of the outcome of selection based entirely on phenotypic considerations, usually "survival of the fittest" kinds of arguments, are common in evolutionary biology. The motivation for using phenotypic arguments is twofold. First, as I have noted several times, selection dynamics can be complex and it would be useful to have shorthand descriptions of their outcome. Second, the genetic basis of the trait of interest is often unknown. Often the lack of knowledge concerning the genetic basis of a trait is used to justify ignoring genetics. I can appreciate the frustration of workers wanting simple answers to complex problems; however, I have never understood the argument that because one does not know the genetic basis of the trait one should pretend to ignore genetics altogether. I say "pretend," because phenotypic approaches (ESS, optimality, and other fitness arguments) do not ignore the genetic basis—most implicitly assume that phenotypes are passed on directly to offspring. In other words, phenotypic arguments implicitly assume that the genetic system is one of asexual cloning. But the organisms of interest are usually sexual and, as we have discovered, sex has dramatic consequences on the character of evolution and the dynamics of selection.

With sex, the genetic system underlying the transmission of phenotypes can impose constraints to their evolution, either through lack of genetic variation or because of "masking" of the heritable capacities. Masking of the heritable capacities refers to a situation in which more adapted types (more adapted according to an ecological or engineering analysis of the capacities upon which $F$-fitness depends) do not outcompete less adapted types. Masking of adaptedness can occur because of genetic constraints, like those stemming from sexual recombination, pleiotropy, or linkage, or as a result of ecological factors, such as frequency-dependent interspecific interactions (Abrams et al. 1993), density (recall the cost of commonness or the cost of rarity), or whether a population is increasing or not (Michod 1995; Byerly and Michod 1991a; Michod 1986). Some examples were encountered in chapters 2 and 3 in our analysis of hypercycles and again in chapter 7 in our study of the role of sex in maintaining species distinct from one another.

With sexual reproduction, the $F$-fitness of one type depends on the heritable capacities of its mate. Sexual selection, as Darwin noted, can drive evolutionary change in a direction of lower viability—which is by commonsense the focus of good adaptive design. In the extreme case of meiotic drive, the selection dynamics, operating on genes as interactors as well as replicators, can select for less adapted organisms to the point of extinction of a local population. These cases are well known, but it is instructive to look to the kind of dynamics involved.

Sexual reproduction introduces an intrinsic nonlinearity into the dynamics of natural selection, since mates must encounter one another for reproduction to occur. The component of $F$-fitness due to the birth process is then a nonlinear function of density, $N_i^2 b_i$, instead of being linear as in equation 8-1. This introduces a cost of rarity into natural selection, in both inter- and intraspecific selection (Hopf and Hopf 1985; Bernstein et al. 1985c), which can result in a decoupling of $F$-fitness from the effects of the heritable capacities. This decoupling describes the situation when the outcome of selection among types cannot be predicted from their individual capacities, and in some cases types with capacities that are intuitively more adaptive lose in selection (Bernstein et al. 1985c; Michod 1986, 1984). The cost of rarity in sexual populations leads to the "survival of the first" paradigm discussed earlier in this chapter, in chapter 1, and with regard to the origin of species in chapter 7.

In a sexual population, the heterozygote with superior fitness cannot win in competition over less fit homozygotes. The classic case of heterozygote superiority, sickle cell anemia (discussed in chapter 7), illustrates a decoupling between "adaptiveness" of phenotypes and the $F$-fitness of corresponding genotypes. An individual homozygous for the sickling trait ($aa$ genotype) might have an expected number of offspring of zero. But the $F$-fitness of the $aa$ genotype will be positive when the heterozygote genotype is increasing in a population.

As already discussed in chapter 7, models of kin selection show the central role of $F$-fitness in natural selection, as opposed to reproductive success of individual organisms. An altruistic genotype may increase in frequency, having higher $F$-fitness within a population, even though the corresponding phenotype (altruism, for example) is less adaptive in the sense that its design decreases its own individual survival and reproduction. Natural selection may cause a decrease in the average individual fitness of a population, especially in cases of frequency-dependent selection.

In conclusion, there is no simple connection between the adaptiveness of the heritable phenotypic properties of an organism (heritable capacities) and the $F$-fitness of the corresponding genotype class, since $F$-fitness depends on the environment, the population, and the genetic system, in addition to its heritable capacities. Of course, there is a connection as specified by the $F$-fitness function. This is the best we can do, and it is quite enough. There is no overall adaptedness, no optimal design of organisms that is maximized by natural selection—even ignoring chance factors such as genetic drift. In general, models of natural selection reveal considerable tension between the intuitive notion of fitness as the overall adaptedness of organisms and the concept of $F$-fitness of genotypic classes of individual units upon which evolution by natural selection ultimately depends.

## ARE ADAPTATION CONCEPTS NECESSARY?

When viewed in the present framework, an adaptation is a heritable capacity, or trait underlying the capacity, that contributes positively to $F$-fitness. Explanatory models of natural selection exhibit the way in which a particular trait contributes positively to fitness, either individual fitness, inclusive fitness, or the $F$-fitness of the class with the trait. Consider the case of altruism. In certain conditions, as when relatives are present, altruism can be adaptive—that is, contribute positively to $F$-fitness of the corresponding genotypic class. As a result, altruism can increase in selective competition with nonaltruism. In other situations, in particular when relatives are not present, altruism is nonadaptive. Since models are usually constructed for the purpose of determining the adaptiveness of a particular trait (or capacity), we referred to heritable capacities as "adaptive capacities" in a previous treatment (Byerly and Michod 1991a,b). We emphasized there the defining property of "a heritable trait, that contributes positively to $F$-fitness." Here I have dropped the term "adaptive."

The term "adaptive capacity" can be confusing for the following reason. Although the heritable capacity (and its underlying trait(s)) can be discussed independently of the environment (even though its value may change with the environment), the adaptiveness of the capacity (in other words, whether it contributes to positive $F$-fitness) depends on the environment. As just mentioned, the adaptiveness of altruism depends upon whether relatives are present. We wish to avoid the following question: Since the adaptiveness of a capacity depends on the environment, must not the adaptive capacities, the components of $\mathbf{a}_i$ in equation 8-2, also be indexed to different environments? A property of an organism (or other unit of selection) can be referred to without having to mention an environment, even though its being adaptive is relative to an environment. We want to be able to refer to the same capacity or trait (and its underlying genetics) being present in different environments, for the obvious reason that we want to know in what environments the trait will increase in frequency, that is, have positive $F$-fitness. This holds even recognizing the complication that, in the developmental process from genes to phenotype, the expression of a trait is a function of both the environment and the genotype. Therefore, to avoid confusion I speak here of "heritable capacities" instead of "adaptive capacities."[16]

Reflection on the meaning of "adaptive capacity," the masking of adaptiveness discussed in the last section, and the problems inherent in overall adaptedness raises questions about general uses of the term "adaptive," and of other commonly found forms of "adapt" in evolutionary biology: "adapted," "adaptedness," and "adaptation." The various forms of "adapt" do not enter into the formalization of natural selection in terms of dynamical equations, such as those given above, which are of the common form typically used in population biology. The concept of adaptation is not represented in dynamical models of natural selection. The various forms of "adapt" are shorthand descriptors for situations in which certain traits contribute positively to $F$-fitness. Although "adaptiveness," "adaptive," and "adaptation" are not needed as technical terms within evolutionary theory, once a dynamical model is formulated, the intuitive concept "adaptive" does play a role in interpreting the process of natural selection. Adaptations are to be explained by their effects in the selection process, effects that contributed positively to $F$-fitness values in the past.

## $F$-FITNESS AND EVOLUTIONARY EXPLANATIONS

Fitness is the most fundamental concept in all of biology, yet it is difficult to define precisely in a way that captures its many uses. None other than R. A. Fisher, one of the founding fathers of the genetic theory of evolution, used it in different senses without notice. Much of the confusion concerning Fisher's celebrated fundamental theorem stems from a subtle shift in the meaning of "fitness" in his work (Price 1972b). The conceptual problems also obtrude into philosophical interpretations of evolutionary explanations. Evolutionary explanations pose a difficult problem for philosophical analysis because of the complex layers of causal relations involved. An organism's phenotype interacts with its environment to determine individual fitness in the sense of the expected number of offspring produced. The phenotype is produced as a result of development during which mutation and selection can produce within-organism change. It is the organism's genotype, not its phenotype, that is passed on to offspring. To make matters more involved, in a sexual species, only a portion of the genotype is passed on, mixing with genes from another organism to determine the offspring's genotype and phenotype. And this assumes that the unit of selection, usually the organism, is well defined. During the passage between evolutionary units, even more complex layers of causality are involved as fitness effects are traded between the different levels.

There are two directions of explanation in the theory of natural selection: explanation *by* $F$-fitness and explanation *of* $F$-fitness (Byerly and Michod 1991a). $F$-fitness may be used to calculate genotype frequency changes and is, in turn, explained by reference to heritable capacities channeled through the genetic and reproductive systems under environmental conditions. $F$-fitness functions exist as both cause and effect at different levels. In individual selection, causes arise at the level of individual organisms through their capacities, but relevant evolutionary effects are at the level of classes of genotypes and their rates of increase. Natural selection operates *at* the level of births and deaths of individual organisms but *on* the level of changing relative frequency of classes of genotypes in populations.

*Explanation by* F-*fitness.*   Values of $F$-fitness determine the evolution of a population in terms of changing genotype frequencies. Fitness here is not an agent or force, nor an interactive cause, but is simply the rate of increase of a type (on a per capita basis), similar to net acceleration in physics, and this use of fitness is similar to the use of acceleration to explain the trajectory of a body. Differences in rates of increase between genotypes explain the changes in genotypic composition of the population and, consequently, the presence (in certain frequency) of heritable capacities and their underlying traits. In this sense, $F$-fitness explains the heritable capacities (that is, the design) of individuals in a population.

*Explanation of* F-*fitness.*   Explanation of $F$-fitness values is causal in a more basic sense: the rate of increase of genotypes is explained on the basis of proximate causes, such as by the interaction of heritable capacities (design) and the environment under the influence of hierarchical organization, the genetic system, and the structure of the population. Heritable capacities represent the components of design in a particular environment that contribute causally to $F$-fitness, again channeled, as it were, through the genetic and reproductive systems. In this sense, the heritable capacities help explain $F$-fitness.

198

That, roughly speaking, heritable capacities explain $F$-fitnesses and $F$-fitnesses explain heritable capacities is possible because of heritability and the repetitiveness of life cycles. Present $F$-fitness values depend in part on heritable capacities (as specified by the $F$-fitness function), and heritable capacities in their present frequencies result from past differences among ancestor genotypes in their $F$-fitnesses. We could say that we explain the process of natural selection by its products (the heritable capacities) as well as by its dynamics ($F$-fitness). However, there is no logical circularity here. We do not, of course, explain the existence of a particular adaptation in an organism by *its* effects, its contribution to $F$-fitness. Rather, an organism has capacities that are adaptive now by virtue of the kinds of effects that the capacities had in ancestors in past environments and the fact that the present environment is similar to past environments. Heritable capacities act as present causes of $F$-fitness values, and are also products of evolutionary causes in the past involving $F$-fitnesses of ancestor genotypes.

Circularity threatens only if "fitness" is viewed as an overall adaptedness propensity of an organism, or other unit of selection. Is the purported fitness as a propensity of an organism in an environment (its overall "adaptedness") supposed to *cause* the organism's expected reproductive success or simply *be* the expected success? Do evolutionary biologists explain the actual "fitness" of an organism, in the sense of its number of offspring, by saying that the organism has a certain "fitness," in the sense of overall fitness propensity? I think not. Instead of appealing to an overall fitness propensity, evolutionary explanations, and the many models upon which they depend, appeal to specific heritable capacities that, when channeled through the reproductive and genetic systems, determine evolutionary success, that is, $F$-fitness values.

Explanations of genetic change require a causal basis for $F$-fitness values. A similar need for a causal basis was responsible for deconstructing the once abstract selection coefficient of population genetics (Michod 1981). The selection coefficient was initially assumed constant and devoid of ecological import. The ecological models of selection mentioned briefly in chapter 1 established a causal basis for the selection parameter of population genetics. Functional explanations provide for a causal basis and are represented by the right-hand side of the $F$-fitness functions (for example, the right-hand side of the equations listed in note 14). A functional explanation shows how heritable capacities, taken in combination with environmental factors and channeled through the genetic system, give rise to the differences in $F$-fitness values of corresponding genotypes. A functional explanation of an adaptation (a heritable capacity that contributes positively

199

to the $F$-fitness of the genotypic class) shows that organisms in a population possess an adaptive trait now because there were differential $F$-fitnesses of corresponding genotypes among their ancestors in the past. Functional explanations in evolution have the general pattern of explaining causes (heritable capacities) in terms of their effects (effects on $F$-fitness values), but they require no future-directed teleology, because of the repetitiveness of the life cycle and the predictability of the environment.

## EXPLAINING FITNESS

Lewontin remarked that evolution by natural selection should explain "fitness" (Lewontin 1978). I have taken the approach that to explain fitness we need to understand (i) how fitness originated in the transition from the non-living to the living, (ii) the role of fitness in the mathematical theory of natural selection, and (iii) how new levels of fitness are created during evolutionary transitions to greater levels of complexity.

Order emerged from the chaos when molecules became capable of self-replication. Once replication began, fitness relations could further direct this transition into the living realm. Although it is possible to refer to the fitness of any physical object in terms of its persistence over time, biological fitness requires a life cycle involving reproduction and death, that is, fitness requires a life history. Fitness relations bind the many components of the life cycle through time, space, and levels of organization.

Mathematical models in population biology make explicit Darwin's general framework for natural selection as expressed in his conditions of variation, heritability, and the struggle to survive. How shall we interpret the mathematical equality between the left- and right-hand sides of the many Darwinian dynamics studied here? The measure of fitness on the left-hand side is set quantitatively equal to a function of parameters and variables on the right-hand side of the equation. The left-hand side of the equation is read as the per capita rate of increase of type $i$. This is the $F$-fitness *value* of a type, often a genotypic class. $F$-fitness values of different genotypes provide a measure of their evolutionary success. The functional relationship on the right-hand side reflects the causal dependence of $F$-fitness on the heritable capacities, $a_i$, and environmental factors, $\mathbf{E}$ (equation 8-2). Included are the systematic interactions among the heritable capacities and the environment. In addition, population structure, organizational structure (in terms of nested hierarchical units), and the genetic and mating systems may enter into the functional form of $F$, affecting the way in which fitness depends on properties of the individual and environment.

To explain the evolutionary success of a type, its $F$-fitness, we look to causal relations between heritable capacities, environmental variables, and genetic and reproductive systems, as we have had many opportunities to do in our studies here of cooperation and conflict in the evolution of individuality and transitions in evolutionary units. Fitness is not an overall property of an organism or of any other unit of selection. Particular traits are explained by representing them in terms of heritable capacities and then by showing how they contribute to $F$-fitness. How do we explain the present adaptiveness of heritable capacities and their underlying traits? Heritable capacities and their underlying traits are both the cause of $F$-fitness differences in the present and an effect of $F$-fitness differences in the past. There is a circularity here but it is causal, not logical, because of the repetitiveness of life cycles.

There is little to recommend the view that fitness increases in evolution, and many reasons to dispute this, most importantly the prevalence of frequency-dependent effects on the components of fitness. There is no overall adaptedness, no overall property of the organism or other unit of selection that is increased during evolution. The mode of explanation in evolutionary biology is in the other direction, to break down vague notions of overall adaptedness into specific components that interact with the environment in the context of the genetic and reproductive systems. Fisher's fundamental theorem (discussed in chapter 4) is the closest one can get to viewing fitness as an increasing quantity in evolution, and it isn't very close. In conceptual terms, Fisher's theorem states that average fitness increases with regard to the environment that existed previously (in the last generation). The problem is that fitness depends upon the environment and the environment is always changing, in large measure because of frequency-dependent responses of other members in the population. Fisher's theorem expresses the commonsense notion of "selection," which is that the selected set should contain more of the property used in selection (if we choose items at the grocery store according to their cost, this should be reflected in the price of items in our grocery bag when we get home). The problem is that there is no choosing agent in evolution, just differential replication and survival of genotypes, and the property under study, fitness, is changing all the time, because of sex, within-level change, frequency-dependent responses of other members of the population, and additional changes of the environment.

Although fitness does not increase in evolution, new levels of fitness can be created. This happens whenever there is a transition in the units of selection. In chapters 5 and 6 we studied how new levels of fitness are created. Fitness connects not only the life cycles of replicating entities at one level, but also units of selection at different levels, as it assembles hierarchically nested units into higher levels of organization. New levels of fitness emerge out of the dynamics of natural selection as fitness is exported from lower to higher levels during the evolution of cooperation. In this theory, fitness is a dynamical concept, both as cause and effect of evolutionary change and by virtue of its own malleability as it is moved between levels. Increments in fitness are traded among levels of selection through the evolution of individually costly yet group-beneficial behaviors by the mechanisms of reciprocation, spatial structure, kin selection, and group selection. If sustained through conflict mediation, this trade results in an increase in the heritability of fitness and individuality at the new higher level. In this way higher levels of fitness—new evolutionary individuals—may emerge in the evolutionary process.

# Supporting Analyses

## STATISTICS OF FITNESS AND SELECTION

For organisms to emerge as a new evolutionary unit of cooperatively interacting cells, heritability of fitness must arise at the cell-group, or organism,
level. This requires a positive regression of organism (adult) fitness on the
heritable propensity in the zygote for cells to cooperate. The regression,
$\text{Reg}[W_i, q_i]$, appears directly in the dynamics of evolution given in equation
4-11. The zygote's propensity to cooperate is determined by the frequency
of cooperative alleles (genotypes in the case of asexual diploidy) in the zygote.

### Haploidy

I first consider haploidy. There are three components to equation 4-11. The
first component is the weighted regression of adult fitness on zygote gene
frequency, which is independent of gene frequency in the case of haploidy
(equation A-1). A graphical interpretation of the regression is given in figure A-1, although the regression need not be positive as drawn. However,
for there to be a meaningful internal equilibrium (equation 5-4), the regression must be positive. The slope of the line in figure A-1 is the regression
given in equation A-1. For the explicit mutation/selection model given in
table 5-3 the regression depends on the generation time $t$, the mutation rate
$\mu$, the level of cooperation among cells, $\beta$, and the rates of replication for
$C$ and $D$ cells.

$$\text{Reg}_{\text{Hap}}[W_i, q_i] \equiv \frac{\text{Cov}_{\text{Hap}}[W_i, q_i]}{\text{Var}_{\text{Hap}}[q_i]} = \frac{q(W_C - \overline{W})(1-q) + (1-q)(W_D - \overline{W})(0-q)}{q(1-q)}$$

$$= W_C - W_D = k_{CC} + \beta \ k_{CC} + k_{DC} - W_D \qquad \text{(A-1)}$$

The components of the regression are the covariance and variance. In haploid populations, the covariance of adult fitness with frequency of the $C$ allele in zygotes is $q(1 - q)(W_C - W_D)$, while the variance of frequency of
the $C$ allele is $q(1 - q)$. Dividing the covariance by the variance in gene fre-

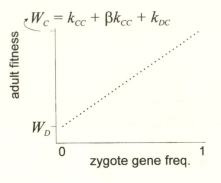

*Figure A-1*

Regression of haploid adult fitness on genotype. See text for explanation.

quency gives equation A-1. The variance in fitness in haploid populations can be shown to be $q(1 - q)(W_C - W_D)^2$.

The second component of equation 4-11 is the gene frequency variance, which is $\mathrm{Var_q}[q_i] = q(1 - q)$. The third component is the average change in gene frequency within organisms, which can be calculated to be

$$E_{\mathbf{Wq}}[\Delta q_i] = -qk_{DC}\left(1+\frac{\beta k_{CC}}{k_{CC}+k_{DC}}\right)\bigg/\overline{W}.$$

After substituting these components into equation 4-11, the total change in gene frequency becomes

$$\Delta q\overline{W} = (k_{CC}+\beta k_{CC}+k_{DC}-W_D)q(1-q)-qk_{DC}\left(1+\frac{\beta k_{CC}}{k_{CC}+k_{DC}}\right).$$

*Sexual Diploidy*

Now I consider diploidy; the additional terms and definitions for the diploid model are given in table A-1. For diploid sex, the regression form of the Price equation 4-11 may be written as

$$\Delta q = \frac{\mathrm{Reg_f}[W_i,q_i]\mathrm{Var_f}[q_i]}{\overline{W}} + E_{\mathbf{Wf}}[\Delta q_i], \qquad (A\text{-}2)$$

with $\mathbf{f} = (f_0, f_1, f_2)$, and $\mathbf{Wf} = (W_0 f_0, W_1 f_1, W_2 f_2)$ used as weighting vectors. For sexual populations the regression and its components are given by

| Term | Meaning |
|------|---------|
| $i, j$ | Subscript for diploid cell type and diploid zygote genotype: $i, j = 0,\ 1,\ 2$ for $DD$, $CD$, $CC$, respectively |
| $f_i$ | Frequency of genotype $i$ in the population |
| $k_{ij}$ | Number of $i$ cells in the adult stage of a $j$-zygote |
| $k_i$ | Total number of cells in adult stage of $i$-zygote after development: $k_i = \sum_j k_{ji}$ |
| $W_i$ | Fitness (expected number gametes) of zygote $i$: $W_i \propto k_i + \beta k_{2i} + d\beta k_{1i}$ |
| $d, h$ | Dominance parameters at level of adult organism and cell, respectively |
| $\overline{W}$ | Average individual fitness: $\overline{W} = \sum_i f_i W_i$ |

*Table A-1*

Additional terms and definitions for the diploid model.

$$\mathrm{Reg}_{\mathrm{Sex}}[W_i, q_i] \equiv \frac{\mathrm{Cov}_{\mathrm{Sex}}[W_i, q_i]}{\mathrm{Var}_{\mathrm{Sex}}[q_i]} = \frac{q(1-q)(-W_0(1-q) + W_1(1-2q) + qW_2)}{\dfrac{q(1-q)}{2}}$$

$$= 2(-W_0(1-q) + W_1(1-2q) + qW_2), \tag{A-3}$$

where $q_i = 0,\ \frac{1}{2},\ 1$ for $i = 0,\ 1,\ 2$.

*Asexual Diploidy*

In the case of asexual diploidy, it is not sufficient to represent evolution in terms of gene frequency, since there is no way to generate the diploid genotype frequencies from gene frequency—no rule akin to the random mating that I am assuming in the case of diploid sex. Consequently, the important variable is genotype frequency itself, and the regressions, one for each diploid genotype, must be calculated with respect to the initial frequencies of the three genotypes in each of the three zygotes, which will, of course, be 0 or 1. Consequently, we must calculate the covariance of adult fitness with genotype of the zygote. The initial frequency of a genotype in a zygote will be 1 for the genotype of the zygote and 0 for the other two genotypes. The vectors of initial frequencies of the three genotypes in each of the three zygotes are given by

$$\mathbf{f}^0 = (1,0,0) \text{ for } DD,\ \mathbf{f}^1 = (0,1,0) \text{ for } CD,\ \mathbf{f}^2 = (0,0,1) \text{ for } CC.$$

205

By using the appropriate vector from this list, the genotypic covariances of fitness can then be calculated directly as follows (the notation $\mathbf{f}^0_i$ means the $i$th element of the vector $\mathbf{f}^0$):

$$\text{Cov}[W_i, \mathbf{f}^0] = \sum_{i=0}^{2} f_i (W_i - \overline{W})(\mathbf{f}^0_i - f_0),$$

$$\text{Cov}[W_i, \mathbf{f}^1] = \sum_{i=0}^{2} f_i (W_i - \overline{W})(\mathbf{f}^1_i - f_1),$$

$$\text{Cov}[W_i, \mathbf{f}^2] = \sum_{i=0}^{2} f_i (W_i - \overline{W})(\mathbf{f}^2_i - f_2).$$

The genotypic covariances in the last set of equations can be simplified to:

$$\text{Cov}[W_i \mathbf{f}^0] = f_0(W_0 - \overline{W}),$$
$$\text{Cov}[W_i \mathbf{f}^1] = f_1(W_1 - \overline{W}), \tag{A-4}$$
$$\text{Cov}[W_i \mathbf{f}^2] = f_2(W_2 - \overline{W}).$$

We also need to calculate the expected within-organism change in genotype frequency. For example, in the covariance equation for genotype $DD$ we need the changes in frequency of the $DD$ genotype from zygote to adult stage in each of the three zygote types. These changes are given in the vector $\Delta f_0$, along with the changes of the other two other genotypes $\Delta f_1$ and $\Delta f_2$, in the equations:

$$\Delta f_0 = \left( \frac{k_{00}}{k_0} - 1, \frac{k_{01}}{k_1} - 0, \frac{k_{02}}{k_2} - 0 \right),$$

$$\Delta f_1 = \left( \frac{k_{10}}{k_0} - 0, \frac{k_{11}}{k_1} - 1, \frac{k_{12}}{k_2} - 0 \right),$$

$$\Delta f_2 = \left( \frac{k_{20}}{k_0} - 0, \frac{k_{21}}{k_1} - 0, \frac{k_{22}}{k_2} - 1 \right).$$

Using these equations, I obtain the three covariance equations for the three zygotes given in equation A-5, which describe evolution in asexual diploid populations:

$$\Delta f_i = \frac{\text{Cov}[W_i, \mathbf{f}^i]}{\overline{W}} + \text{E}[\Delta f_i], \quad i = 0,1,2. \tag{A-5}$$

In the case of asexual diploidy we may consider two sets of statistics: the first calculated with respect to zygote gene frequency (as we have done for sexual diploidy) and the second calculated with regard to zygote genotype frequency as we did in the derivation of equation A-5. In the case of asexual diploidy, the statistics calculated with respect to gene frequency bear no necessary relation to the evolutionary dynamics, while the statistics calculated with regard to zygote genotype frequency enter into equation A-5 directly. When statistics are calculated with respect to zygote genotype frequency, there are three statistics, one for each genotype. For the purpose of comparison with sexual diploidy (in which there is just a single statistic), I average these three statistics in the figures that follow. For example, for comparisons of the genotypic regressions of fitness on frequency, I use the average of the three regressions corresponding to the three covariances given in equation A-5, using as weights the three genotype frequencies in the population $f_i$ for $i = 0,1,2$. In a similar fashion I average the genotypic covariances and variances. The needed genotypic variances in individual genotype frequency can be calculated directly and are given next:

$$\text{Var}[\mathbf{f}^0] = \sum_{i=0}^{2} f_i(\mathbf{f}^0{}_i - f_0)^2 = f_0(1 - f_0),$$

$$\text{Var}[\mathbf{f}^1] = \sum_{i=0}^{2} f_i(\mathbf{f}^1{}_i - f_1)^2 = f_1(1 - f_1),$$

$$\text{Var}[\mathbf{f}^2] = \sum_{i=0}^{2} f_i(\mathbf{f}^2{}_i - f_2)^2 = f_2(1 - f_2).$$

The average genotypic regression and average genotypic covariance for asexual populations simplify as

$$\overline{\text{Reg}_{\text{Asex}}} = \sum_{i=0}^{2} \frac{f_i}{1 - f_i}(W_i - \overline{W}),$$

$$\overline{\text{Cov}_{\text{Asex}}} = \sum_{i=0}^{2} \text{Cov}[W_i, \mathbf{f}^i] f_i = \sum_{i=0}^{2} f_i^2(W_i - \overline{W}). \tag{A-6}$$

Sex dramatically changes how a genotype's covariance of fitness with zygote frequency contributes to the overall covariance of fitness in the population. I now express the sexual covariance of genotypic fitness on individual *gene* frequency (given in the numerator of equation A-3) in terms of the genotypic covariances of equation A-4:

$$\text{Cov}_{\text{Sex}} = \sum_{i=0}^{2} \text{Cov}[W_i, \mathbf{f}^i](q_i - q). \tag{A-7}$$

*Statistics of Fitness*

Without sex, a genotype's covariance is simply averaged using the population genotype frequencies as in the development of equation A-6. Of course, the genotypic covariances themselves may be positive or negative depending on the model, as we will see. Nevertheless, without sex the sign of each is preserved in its contribution to the average.

All this changes with sex. From equation A-7, we see that the contribution of the genotypic covariances to the sexual covariance depends on the relationship of the current population gene frequency, $q$, and the individual gene frequency, $q_i = 0, 1/2, 1$ for $i = 0, 1, 2$. The contribution of the $DD$ genotypic covariance $Cov[W_i, \mathbf{f}^0]$ to the sexual covariance is always negative, while the contribution of the $CC$ genotypic covariance $Cov[W_i, \mathbf{f}^2]$ is always positive. However, the strength of the contribution of each depends on the current population gene frequency. For example, if the population gene frequency is close to 1, then the $DD$ genotype will contribute a large amount since $(q_0 - q) \approx 0 - 1 = -1$, while the $CC$ genotype will contribute little to the population covariance, since $(q_2 - q) \approx 1 - 1 = 0$. The contribution of the heterozygote $CD$ covariance depends on whether the current gene frequency is greater than or less than $1/2$ and the degree of the difference from $1/2$.

I now shall consider further the statistics of fitness and selection for sexual and asexual diploidy and for haploidy, for the parameter value cases considered in figures 5-3 through 5-5 in the text. Regressions of adult (individual) fitness on zygote frequency of cooperation are graphed in figure A-2 for additive mutations as a function of the development time (panels A and B), the mutation rate (panels C and D), the benefit to organisms of cooperation (panel E), and the advantage to cells of defecting (panel F). The corresponding variances of adult fitness are graphed in figure A-3. To understand the regressions it is necessary to understand something about the nature of the two kinds of equilibria under diploidy: mutation-selection balance and heterozygote superiority (figure A-4). The variances of zygote frequency are graphed in figure A-5. The regression of adult fitness on zygote frequency of cooperation for recessive mutations is graphed in figure A-6. For reasons of space, the variances of adult fitness and zygote frequency for recessive mutations are not given here (they are available from the author). The figures for diploidy corresponding to figures 5-3–5-5 have been studied in Michod (1997a, figs. 8–11). For comparison purposes, the panels of all of these figures correspond to the same sets of parameter values (figures 5-3–5-5, A-2, A-3, A-4, and A-6 here and figures 8–12 of Michod 1997a).

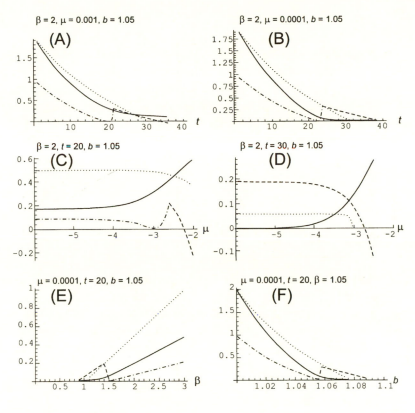

*Figure A-2*

Regression of fitness on individual frequency at equilibrium for mutations with intermediate dominance. This fitness regression is a component of the first component of the Price equations 5-2, A-2, and A-5. The parameter values are given above each panel. The horizontal axis varies between panels. Haploidy: dotted lines; diploid sex: solid lines; diploid asexuality: dashed lines (first equilibrium) and dashed-dotted lines (second equilibrium). Regressions are shown only for locally stable equilibria described in Michod 1997a, especially figure 8 of that paper.

Before considering the results given in these figures, there are several technical points that need to be understood. The regression for haploidy does not depend on gene or genotype frequency (equation A-1); however, the diploid regressions do. For this reason, the diploid statistics are calculated at the stable-equilibrium genotype frequencies determined by the parameter values. The frequency of cooperation, average fitness, and covari-

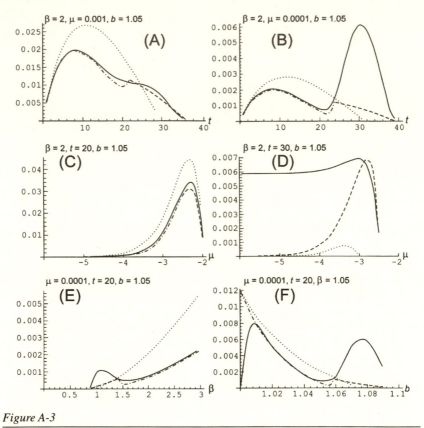

*Figure A-3*

Variance in fitness at equilibrium for mutations with intermediate dominance. This variance is a component of the first component of the Price equations 5-2, A-2, and A-5. Legend the same as for figure A-2.

ance of fitness at these equilibria are shown in Michod 1997a, figures 8–11. As explained in more detail in that reference, for asexual diploidy there is an apparent discontinuity in the graphs at the point when the two equilibria exchange stability. In fact there are really two curves for asexual diploidy corresponding to the two equilibria given there in equations B3 and B4. However, the statistics are not graphed for the unstable or biologically unrealistic portions of the equilibrium curves.

For asexual diploids, I graph the average of the statistics (regressions and variances in zygote frequency) using the zygote genotype frequencies, since it is these statistics that determine the evolution of the system (using the corresponding regression form of equation A-5). I have also

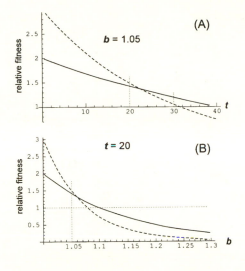

*Figure A-4*

Relative fitness of genotypes at equilibria studied. $W_{CD}$, solid curve, $W_{CC} = W_C$ dashed curve. $W_{DD} = W_D = 1$. In panel (A), relative fitness is plotted as a function of development time for $b = 1.05$. In panel (B), relative fitness is plotted as a function of advantage for defection for $t = 20$. The figure is based on the model given in table A-1 using a mutation selection model similar to table 5-3 but for diploidy (Michod, table A1 1997a). Other parameter values for both panels are $\beta = 2$, $c = 1$, $\mu = 0$, and $h = d = 0.5$. The vertical dotted lines indicate values of parameters assumed in figures A-2, A-3, A-5, and A-6.

considered the regression and variances using zygote gene frequency, and there is little qualitative difference in the behavior of the curves, except that the regressions are larger in magnitude and the variances are smaller. The variance in fitness graphed in figure A-3 is unaffected by these considerations.

In the case of recessive mutations, the first asexual diploid equilibrium is exactly the same as the haploid equilibrium. In this case, $k_{01} = k_{DC}, k_{11} = k_{CC}$, and $k_1 = k_C$ in table A-1, and so $W_C = W_1$ (using also table 5-2 and Michod 1997a, table A1). As a consequence, the equilibria given for haploidy (equation 5-4) and asexual diploidy (Michod 1997a, eq. B3) are the same. In other words the frequency of $CD$ cells at the diploid equilibrium is the same as the frequency of $C$ cells at the haploid equilibrium. So the frequency of the $C$ allele at the diploid equilibrium is exactly 1/2 the haploid value (Michod 1997a, fig. 9). However, the expression of cooperativ-

211

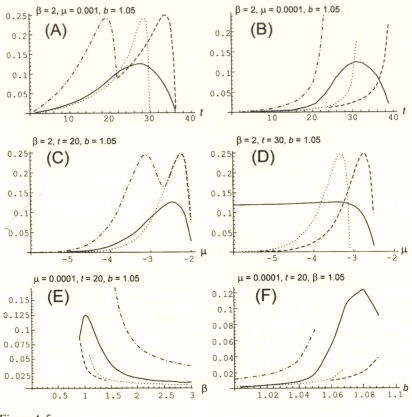

*Figure A-5*

Variance in zygote frequency at equilibrium for mutations with intermediate dominance. Legend the same as for figure A-2.

ity is exactly the same, since the *C* allele is dominant to the *D* allele in the heterozygote. If we use the zygote gene frequency allele for the asexual diploid statistics, it can be shown that the covariance under diploidy is one-half the covariance under haploidy. However, the variance in individual gene frequency is one-fourth the variance under haploidy. So the diploid regression is twice the haploid regression. However, if we use zygote genotype frequencies for the asexual diploid statistics, the asexual diploid variance equals the variance of zygote gene frequency under haploidy. The covariance and hence the regression (since the variances are the same) under diploidy deviate from the haploid statistics by a (usually small) factor

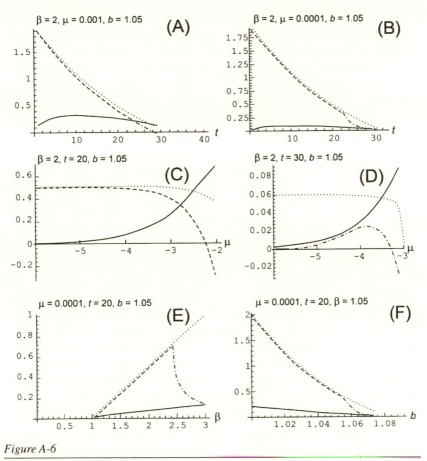

*Figure A-6*

Regression of fitness on individual frequency at equilibrium for recessive mutations. Legend the same as for figure A-2 except the equilibria correspond to figure 9 of Michod 1997a.

$(2f_1 - 1)$. This is close to 1 (i.e., no deviation so haploidy and asexual diploidy give the same statistics) if $f_1$ is close to 1, which is the case for small mutation rates (see Michod 1997a, fig. 9).

Now we consider the results. As shown in panels A and B of figure A-2, as the time available for development increases, or alternatively as adult size increases, the regression of adult fitness decreases. The regressions remain positive—except for asexual diploids in which the regression becomes negative for large organisms and high mutation rates (see figure

213

A-2, panels A, C, and D). The regressions decrease with the strength of within-organism selection as determined by the replication advantage of defecting cells (panel F). The regressions increase with the magnitude of benefit of cooperation (panel E). So far, the behavior of the regression curves as a function of the parameters is as we expect—the regression of adult fitness on propensity to cooperate increases with the benefit of cooperation and decreases with the magnitude of within-individual variation and selection (increased $t$ and $b$).

However, the relation between the fitness regressions and the mutation rate is somewhat surprising (panels C and D of figure A-2 and of figure A-6 above). For haploidy and asexual diploidy, the regressions decrease (and eventually become negative for asexual diploidy) as the mutation rate increases. This makes sense, because deleterious mutation should reduce the mapping of zygote frequency on adult phenotype. However, for sexual organisms, the regressions increase in magnitude as the mutation rate increases. How can this be?

At equilibrium the two components on the right-hand side of the Price regression equation A-2 must equal one another. As the mutation rate increases, the within-organism change must increase, regardless of the breeding and reproductive system. Recall that the within-organism change is measured by the second component of the right-hand side of equation A-2. As the within-organism change increases, the first component on the right-hand side of equation A-2 must also increase in magnitude so as to maintain an equilibrium. However, the first component is a product of two terms: the regression of adult fitness on zygote gene frequency (graphed in figure A-2) and the variance in zygote gene frequency (graphed in figure A-5 above). The increasing within-organism change (as the mutation rate increases in panels C and D of figure A-2 and of figure A-6 above) is compensated for in one of two ways, either by an increase in the variance in gene frequency as in asexual diploids and haploids, or by an increase in the regression of fitness on zygote gene frequency as in sexual diploids.

To understand why this occurs, it is necessary to understand some characteristics of the equilibria possible in the models studied here and in Michod (1997a). There are three kinds of equilibria in the models: fixation of either the $C$ or $D$ allele, mutation selection balance, and heterozygote superiority in the case of diploidy. Within-organism mutation from $C$ to $D$ can only make matters worse for cooperation, so I begin by considering a simpler situation in which there is no mutation and no within-organism selection. In this case, there is typically fixation of one of the genotypes.

| Genotype | Relative Fitness |
|----------|------------------|
| $CC$ | $2^{(1-b)ct}(1+\beta)$ |
| $CD$ | $2^{(1-b)cht}(1+\beta d)$ |
| $DD$ or $D$ | $1$ |
| $C$ | $2^{(1-b)ct}(1+\beta)$ |

*Table A-2*

Relative fitness with no mutation.

However, in the case of sexual diploidy an internal equilibrium is possible due to heterozygote superiority. Consider the relative genotypic fitnesses in table A-2, assuming no mutation in the model of selection given in table A-1 using a mutation selection model similar to table 5-3 but for diploidy (Michod 1997a, table A1). Assuming $b > 1$, the relative fitness of $CC$ and $C$ genotypes decreases with $t$ and increases with $\beta$ (table A-2). The same is true of the relative fitness of $CD$ heterozygotes, except the effects of $t$ and $\beta$ are moderated by the dominance parameters. The relative fitness of both $CC$ and $CD$ decreases with the magnitude of $b$. In table A-2 the relative fitness of $CC$ and $CD$ genotypes as a function of development time, $t$, is plotted for parameter sets including those used in figures A-2, A-3, A-5, and A-6 below.

For lower values of development time, the fitness ranking is $W_{CC} > W_{CD} > W_{DD}$, and the $C$ allele reaches fixation under both diploidy and haploidy ($t < 22$) (consider panel (A) of figure A-4). As $t$ increases past about $t = 22$, the heterozygote becomes superior; the ranking is first $W_{CD} > W_{CC} > W_{DD}$ and then $W_{CD} > W_{DD} > W_{CC}$. Eventually ($t > 40$) the $DD$ homozygote becomes most fit ($W_{DD} > W_{CD} > W_{CC}$) and the $D$ allele becomes fixed. Similar effects occur as $b$ increases and $t$ is fixed, as shown in panel (B) of figure A-4.

In figure A-3 the variance in fitness is graphed for the same set of parameters and conditions as in figure A-2. For sexual diploids, as the mutation rate declines the variance in fitness approaches zero in panel C but is finite and significant in panel D, while for asexuals the variance in fitness approaches zero in both panels as the mutation rate declines. The only difference between the two panels is that $t = 20$ in panel (C) and $t = 30$ in panel D. Recall from figure A-4 that the ranking of fitness for no mutation changes from $W_{CC} > W_{CD} > W_{DD}$ (fixation of $C$ allele) to $W_{CD} > W_{CC} > W_{DD}$

215

(heterozygote superiority). Consequently, for sexuals the variance in fitness is finite even for small $\mu$ in panel D, because of an internal equilibrium caused by heterozygote superiority (very little of this variance is heritable, however, as shown in panel D of figure A-2). For sexual diploids, the fitness regressions increase with increasing within-organism change in both panels C and D of figure A-2. In panel C the regression increases from a finite value, while in the case of heterozygote superiority in panel D, the regression increases from a near-zero value.

Many mutations are recessive, or nearly so. In figure A-6, the regression statistics are graphed for recessive mutations for the same set of conditions as considered in figure A-2. The corresponding equilibria in gene frequency are given in Michod 1997a, fig. 9. The regressions of fitness on zygote frequency in sexual organisms are dramatically affected by the recessivity of mutations and are generally significantly lower than their haploid and asexual counterparts. As development time increases the regressions decrease for haploidy and asexual diploidy in both panels A and B. However, for sex the regressions peak for intermediate values of development time.

Increasing the mutation rate causes a decline of the regressions for haploidy and asexual diploidy but cause the regressions for sex to increase. The behavior for sexual diploidy is similar to the behavior shown in figure A-2, except there is no longer a heterozygote superiority equilibrium in the case of recessive mutations. In panel (C), we see how, as already mentioned, the regressions for the haploidy and asexual diploidy approach one another as the mutation rate declines. Recalling that dashed curves correspond to the first asexual diploid equilibrium (Michod 1997a, eq. B3) and dashed-dotted curves correspond to the second asexual diploid equilibrium (Michod 1997a, eq. B4), we see that only the first equilibrium is biologically valid for the set of parameter values considered in panel C and only the second equilibrium is valid for the set of parameters considered in panel D (figure A-6). In panel E we see that the regressions generally increase with $\beta$, the advantage of cooperation, except for asexual diploidy in a narrow region of values. In panel F the regressions decline with an increasing advantage to cells defecting.

In conclusion, with sex, as the mutation rate increases (and so the amount of within-organism change), more and more of the variance in fitness is heritable, regardless of whether the mutations are additive (figure A-2) or recessive (figure A-6) and regardless of whether the equilibrium involves heterozygote superiority or not (panels D or C, respectively, of figures A-2–A-6). The effect of sex in maintaining a high mapping of adult .

fitness on zygote genotype under the challenging conditions of high mutation is robust in a variety of genetic situations. As the mutation rate increases, the cooperative alleles inherited in the zygote must be converted to noncooperative alleles in the adult. This is true regardless of whether there is sex or not. However, without sex, there is nothing else involved in the transition to the offspring of the next generation and no possibility of undoing the effects of within-organism change toward defection. However, with sex, there is the possibility that a mate will reintroduce cooperative alleles into the offspring.

## Equilibria for Modifier Model ($G = 0$)

There are four dynamical equations (equations 6-1 and 6-2), and the equilibria described in table 6-2, and given mathematically in table A-3, are obtained by setting them equal to zero. The modifier allele may either be absent (table A-3, eq. 3) or fixed (table A-3, eq. 4) in the population, because there is no mutation affecting this allele.

There are four state variables corresponding to the frequencies of the four gamete types, but only three are independent, since they must sum to one. Consequently, there are three eigenvalues determining the stability of each equilibrium, given as table A-4. Some of these eigenvalues depend on whether reproduction is sexual or asexual and the value of the recombination rate $r$. As seen in table A-4, the eigenvalues are ratios of fitnesses multiplied by cell type frequencies ($C$ or $D$). Again, these eigenvalues can be expressed as ratios of fitnesses multiplied by heritability of fitness at the organism level. Equilibrium 2 (no cooperation, modifier allele fixed) is never stable, as I assume that the modifiers of within-organism change accrue some cost. Because of the cost of the modifier, $W_4 > W_3$, whatever the values of the parameters, and so $\lambda_{23} > 1$. This means that the evolution of functions to protect the integrity of the organism is not possible, if there is no conflict among the cells in the first place. It is conflict itself that sets the stage for the evolution of individuality. The eigenvalues of the other three equilibria, in the case of asexual reproduction, indicate that two or three equilibria cannot be stable at the same time. In the case of sexual reproduction I have not been able to determine this analytically, because I have not obtained simple expressions for the eigenvalues ($\lambda_{31}$, $\lambda_{32}$, $\lambda_{41}$, and $\lambda_{43}$ in table A-4). As discussed in chapter 6. We have discovered regions where equilibria 1 and 4 are both stable and regions where equilibria 3 and 4 are both unstable (Michod and Roze 1998).

217

| Eq. | Genotype Frequencies | Allele Frequencies | Description of Loci | Interpretation |
|---|---|---|---|---|
| 1 | $x_1 = 0$, $x_2 = 0$, $x_3 = 0$, $x_4 = 1$ | $q = 0, s = 0$ | No cooperation; no modifier | Single cells, no organism |
| 2 | $x_1 = 0$, $x_2 = 0$, $x_3 = 1$, $x_4 = 0$ | $q = 0, s = 1$ | Cooperation; modifier fixed | Not of biological interest |
| 3 | $x_1 = 0$ <br> $x_2 = \dfrac{W_2 \dfrac{k_{22}}{k_2} - W_4}{W_2 - W_4}$ <br> $x_3 = 0$ <br> $x_4 = \dfrac{k_{42} W_2}{k_2(W_2 - W_4)}$ | $q = \dfrac{W_2 \dfrac{k_{22}}{k_2} - W_4}{W_2 - W_4}$ <br> $s = 0$ | Polymorphic for cooperation and defection; no modifier | Group of cooperating cells: no higher-level functions |
| 4 | $x_1 = \dfrac{W_1 \dfrac{K_{11}}{K_1} - W_3}{W_1 - W_3}$ <br> $x_2 = 0$ <br> $x_3 = \dfrac{K_{31} W_1}{K_1(W_1 - W_3)}$ <br> $x_4 = 0$ | $q = \dfrac{W_1 \dfrac{K_{11}}{K_1} - W_3}{W_1 - W_3}$ <br> $s = 1$ | Polymorphic for cooperation and defection; modifier fixed | Organism: integrated group of cooperating cells with higher-level function mediating within-organism conflict |

*Table A-3*

Equilibria of Two-Locus Modifier Model. The variables $q$ and $s$ are the frequencies of the $C$ and $M$ alleles in the population. Linkage equilibrium assumed.

## COST OF SEX IN DIPLOIDS

In table A-5, I consider a mutation in a sexually reproducing self-fertile diploid hermaphrodite species, *G*, that causes a female to produce diploid offspring asexually. This table is constructed in a similar fashion to table 7-1 in the text and follows the construction in Fisher's (1941) table about the evolution of selfing. The change in frequency of the *G* allele is zero in this case; there is no cost of sex for self-fertile hermaphrodites.

|  | Equilibrium 1 | | | Equilibrium 2 | | |
|---|---|---|---|---|---|---|
|  | $\lambda_{11}$ | $\lambda_{12}$ | $\lambda_{13}$ | $\lambda_{21}$ | $\lambda_{22}$ | $\lambda_{23}$ |
| **Asexual** | $\dfrac{W_1 \dfrac{K_{11}}{K_1}}{W_4}$ | $\dfrac{W_2 \dfrac{k_{22}}{k_2}}{W_4}$ | $\dfrac{W_3}{W_4}$ | $\dfrac{W_1 \dfrac{K_{11}}{K_1}}{W_3}$ | $\dfrac{W_2 \dfrac{k_{22}}{k_2}}{W_3}$ | $\dfrac{W_4}{W_3}$ |
| **Sexual** | $\dfrac{(1-r)W_1 \dfrac{K_{11}}{K_1}}{W_4}$ | $\dfrac{W_2 \dfrac{k_{22}}{k_2}}{W_4}$ | $\dfrac{W_3}{W_4}$ | $\dfrac{W_1 \dfrac{K_{11}}{K_1}}{W_3}$ | $\dfrac{(1-r)W_2 \dfrac{k_{22}}{k_2}}{W_3}$ | $\dfrac{W_4}{W_3}$ |

|  | Equilibrium 3 | | | Equilibrium 4 | | |
|---|---|---|---|---|---|---|
|  | $\lambda_{31}$ | $\lambda_{32}$ | $\lambda_{33}$ | $\lambda_{41}$ | $\lambda_{42}$ | $\lambda_{43}$ |
| **Asexual** | $\dfrac{W_1 \dfrac{K_{11}}{K_1}}{W_2 \dfrac{k_{22}}{k_2}}$ | $\dfrac{W_3}{W_2 \dfrac{k_{22}}{k_2}}$ | $\dfrac{W_4}{W_2 \dfrac{k_{22}}{k_2}}$ | $\dfrac{W_2 \dfrac{k_{22}}{k_2}}{W_1 \dfrac{K_{11}}{K_1}}$ | $\dfrac{W_3}{W_1 \dfrac{K_{11}}{K_1}}$ | $\dfrac{W_4}{W_1 \dfrac{K_{11}}{K_1}}$ |
| **Sexual** | $\dfrac{a_1 + \sqrt{b_1}}{c_1}$ | $\dfrac{a_1 + \sqrt{b_1}}{c_1}$ | $\dfrac{W_4}{W_2 \dfrac{k_{22}}{k_2}}$ | $\dfrac{a_2 - \sqrt{b_2}}{c_2}$ | $\dfrac{W_3}{W_1 \dfrac{K_{11}}{K_1}}$ | $\dfrac{a_2 + \sqrt{b_2}}{c_2}$ |

*Table A-4*

Eigenvalues for asexual and sexual reproduction. For each of the four equilibria described in table A-3, the eigenvalues are given for both asexual and sexual reproduction. Although the equilibrium frequencies do not depend upon the reproductive system, the stability does. As explained in chapter 6, these eigenvalues can be rewritten as the product of fitness and heritability at the organism level. In the case of sexual reproduction $a_1, b_1, c_1$ and $a_2, b_2, c_2$ are complicated terms available from the author by request.

In table A-6, the sexual mutation, $G$, is assumed to arise in a sexually reproducing diploid species with separate male and female sexes. This table is constructed in a similar fashion to table A-5. In table A-6 I assume that the sexual genotypes are separated into males and females with an equal sex ratio. Genotypes $gg$ are sexual, $GG$ asexual, and $Gg$ sexual with probability $h$ and asexual with probability $1 - h$. The change in frequency is given in the legend to table A-6. It can be shown that as the frequency of the $G$ al-

| | Initial | gg | Gg | GG |
|---|---|---|---|---|
| gg♀♀ | $\frac{1}{2}P$ | $\frac{1}{2}PP_g$ | $\frac{1}{2}PP_G$ | 0 |
| gg♂♂ | $\frac{1}{2}P$ | $\frac{1}{2}Po_g$ | $\frac{1}{2}Po_G$ | 0 |
| Gg♀♀ | $Qh$ | $\frac{1}{2}QhP_g$ | $\frac{1}{2}QhP_g + \frac{1}{2}QhP_G$ | $\frac{1}{2}QhP_G$ |
| Gg♂♂ | $Qh$ | $\frac{1}{2}Qho_g$ | $\frac{1}{2}Qho_g + \frac{1}{2}Qho_G$ | $\frac{1}{2}Qho_G$ |
| Gg♀ | $2Q-2Qh$ | 0 | $2Q-2Qh$ | 0 |
| GG♀ | $R$ | 0 | 0 | $R$ |

*Table A-5*

No cost of sex in a diploid hermaphrodite Mendelian population. The initial frequencies of genotypes gg, Gg, and GG are $P$, $2Q$, and $R$ and are divided according to male (♂♂) and female (♀♀) function, in the case of hermaphrodites, and asexual females (♀), as indicated in the first column of the table. The next three columns in the table give the frequencies of genotypes in the next generation of the offspring of the parents in the initial generation. The variables $p_g$ and $o_g$ are the frequencies of the g allele in sperm and eggs, respectively, given by $p_g = o_g = (P + Qh)/(P + 2hQ)$ and $p_G = o_G = Qh/(P + 2hQ)$.

| | Initial | gg | Gg | GG |
|---|---|---|---|---|
| gg♀♀ | $\frac{1}{2}P$ | $\frac{1}{4}PP_g$ | $\frac{1}{4}PP_G$ | 0 |
| gg♂♂ | $\frac{1}{2}P$ | $\frac{1}{4}Po_g$ | $\frac{1}{4}Po_G$ | 0 |
| Gg♀♀ | $Qh$ | $\frac{1}{4}QhP_g$ | $\frac{1}{4}QhP_g + \frac{1}{4}QhP_G$ | $\frac{1}{4}QhP_G$ |
| Gg♂♂ | $Qh$ | $\frac{1}{4}Qho_g$ | $\frac{1}{4}Qho_g + \frac{1}{4}Qho_G$ | $\frac{1}{4}Qho_G$ |
| Gg♀ | $2Q-2Qh$ | 0 | $2Q-2Qh$ | 0 |
| GG♀ | $R$ | 0 | 0 | $R$ |

*Table A-6*

Cost of sex in a diploid Mendelian population. In the case of sexual parents, only offspring of the same gender are counted so as not to count sexually produced offspring twice. The legend is the same as for table A-5, except the "frequencies" in the last three columns no longer sum to unity and are understood to be relative frequencies. The change in gene frequency for the asexual allele $G$, in frequency $p$, is $\Delta p = (PQ + hRQ + PR - hPQ)/(1 + R + 2Q(1 - h))$. Note that the full dynamical system involves all three genotypic frequencies.

lele gets small, the frequency of the $G$ allele increases by a factor $2(1 - h)$, which is a doubling if the asexual mutation is dominant ($h = 0$). This demonstrates the cost of sex in a diploid Mendelian species when the mutation has some penetrance ($h > 0$). Of course, recessive asexual mutation ($h = 1$) are neutral when rare.

My principal reason for analyzing the cost of sex is to show that although individual fitness is the same for asexual and sexual females (expected number of surviving offspring is equal), asexual females may increase nevertheless. One might point out correctly that the number of grandchildren is different for asexual and sexual females (when there are separate sexes). Counting grandchildren is another way of including the effects of transmission and the reproductive system, the importance of which for evolution is my main point.

# Fitness Phrases

THERE ARE many different legitimate senses of "fitness." In the present book, I take care to clarify the sense being used when this is not apparent from the context. In table B-1, I give a sample from the literature of quotations involving "fitness." I have tried to include the basic uses, but, of course, no list could be complete. Other than in the phrase "survival of the fittest," Darwin rarely used the word "fitness," so I have included as entries 4 and 5 in the table some quotations from Darwin's discussions of natural selection. For each of the phrases in table B-1, I consider certain basic distinctions: whether the unit of selection is the gene, the individual organism, or the population, whether fitness is being used as a cause of evolution or as an effect of natural selection, and whether fitness is being used in the sense of good design or as an operational measure of success. I also note if the author is using fitness in the sense of overall adaptedness. Sometimes the author's meaning in these regards is not clear, so I have taken the liberty of interpreting the author as best I could.

| Fitness Phrase | Level | Cause or Effect | Success or Design |
|---|---|---|---|
| 1. Survival of the fittest | Individual | Cause | Design |
| 2. Progressive increases of fitness of each species of organism | Population | Effect | Design |
| 3. Fitness is the expected reproductive success of a genotype | Individual | Cause | Success |
| 4. Under nature, the slightest difference of structure or constitution may well turn the nicely balanced scale in the struggle for life, and so be preserved · | | | |
| 5. Each new form will tend in a fully stocked country to take the place of, and finally to exterminate, its own less improved parent or other less favored forms with which it comes into competition | | | |
| 6. Fitness is not an explanation, it is only a *description*. It quantifies the biological properties that lead to shifts in genotype distributions; it does not cause the shifts | Individual | Effect | Success |
| 7. [The Malthusian parameter] *m* measures fitness by the objective fact of representation in future generations | Population | Effect | Success |
| 8. The rate of increase of the average fitness of the population is equal to the genetic variance of fitness of that population | Population | Both | Success |
| 9. We wish to define a fitness measure for the age-structured case which determines the equilibrium gene frequencies in the same way as the $W_{ij}$ in the discrete-generation case . . . | Individual | Cause | Success |
| 10. [Fitness is the] ability of organism to survive and reproduce in an environment | Overall adaptedness | Effect | Design |
| 11. [Fitness is the] conformity between the organism and its environment | Overall adaptedness | Effect | Design |
| 12. [Fitness is the] "biological property" of an organism in an environment measured by its expected fitness value | Overall adaptedness | Effect | Design |

*Table B-1*

Fitness Phrases.

223

| Fitness Phrase | Level | Cause or Effect | Success or Design |
|---|---|---|---|
| 13. Fitter traits increase in frequency, and less fit traits decline | Individual | Cause | Design |
| 14. Evolution by natural selection should explain fitness | Individual | Effect | Both |
| 15. Fitness cannot be a property of the indidual but only of the information that it carries and that, by reproduction, it transmits to the next generation | Information | Effect | Success |
| 16. Fitness is not a physical property, although each organism's fitness consists in the constellation of physical properties it possesses | Overall adaptedness | Both | Design |
| 17. Fitness is an explanatory concept referring to a property which is causally relevant to reproductive success | Overall adaptedness | Both | Design |
| 18. A feature of all types of adaptation . . . is that the trait values change in a direction which increases fitness | Individual | Effect | Design |
| 19. It may therefore be worth while to examine in detail a model involving powerful selection, in which the fitness of the species as a whole, judged by external criteria, is entirely inoperative | Population | Both | Design |
| 20. When asking about the dynamics of selfish genetic elements . . . it is inappropriate to assume that "individual fitness" is ever maximized | Individual | Either | Either |
| 21. In the ESS population, fitness is maximized only in the sense that mutants not playing the ESS do worse | Population | Effect | Success |
| 22. Fitter traits increase in frequency and less fit traits decline | Individual | Cause | Design |
| 23. Individuals heterozygous for one or more proteins generally develop less fluctuating asymmetry and have higher fitness as | Individual | Effect | Success |

*Table B-1*

Fitness Phrases (*continued*).

| Fitness Phrase | Level | Cause or Effect | Success or Design |
|---|---|---|---|
| measured by growth, fecundity, and various measures of physiological performance | | | |
| 24. The fitness of an evolutionary entity, when the entity is defined by a set of attributes determining how it interacts with its environment, is manifested as persistence of those attributes | Varied | Effect | Success |
| 25. Fitness is a consequence of the adaptedness of an entity to its environment | Varied | Effect | Success |
| 26. We must never forget that adaptive processes are manifest only at the level of a population reproducing through time, and individual fitnesses are only one factor of many influencing the processes. . . . Obviously, the course of adaptation cannot be predicted from even a total knowledge of how individual genotypes respond to environments to produce fitness phenotypes | Individual | Cause | Success |
| 27. The only feature fitness has that could be used in a general characterization of it is its effect on reproduction | Overall adaptedness | Both | Design |
| 28. If a set of traits is more fit than another set, then an imaginary gene coding for it spreads faster in the population than does an alternative allele coding for the other set, the rate of increase of the gene being the measure of fitness | Gene | Both | Success |

*Table B-1*

Fitness Phrases (*continued*). *Sources:* 1. Darwin 1872, Herbert Spencer and others; 2. Darwin 1872, p. 41; 3. standard definition; 4. Darwin 1872, p. 83; 5. Darwin 1972, p. 172; 6. Endler 1986, p. 33; 7. Fisher 1958, p. 37; 8. Fisher 1941; 9. Charlesworth 1980, p. 131; 10. Dobzhansky 1969; 11. Pianka 1978; 12. Brandon 1980; 13. Sober 1993, p. 66; 14. Lewontin 1978; 15. Gliddon and Gouyon 1989; 16. Sober 1984, p. 376; 17. Lennox 1991; 18. Abrams et al. 1993; 19. Fisher 1941; 20. Marrow et al. 1996; 21. Parker 1990; 22. Sober 1993, p. 66; 23. Møller 1993; 24. Burns 1992; 25. Burns 1992; 26. Templeton 1982; 27. Rosenberg 1983, p. 463; 28. Fagerstrom 1992.

# Notation

I HAVE TRIED to be consistent in my use of notation: for example, $W$ refers to absolute individual fitness (expected reproductive success); $F$ refers to Fisherian fitness—except when they don't! In some cases, I have redefined terms and variables (even within a single chapter) to be more consistent with usage in the literature. For example, in addition to referring to Fisherian fitness, $F$ refers to Wright's "fitness function" and his $F_{ST}$ statistic. The variable $R$ means resources in chapters 2 and 3, relatedness in discussions of kin selection, genotype frequency in Fisher's model for the average effect of a gene substitution, and fitness in the prisoner's dilemma game. Depending on context the variable $b$ refers to birth, the benefits of altruism to the individual, or the benefits of defection to the cell. I think the context and meaning is clear in most cases, but I apologize for any confusion these changes in meaning may create.

Throughout the book the following conventions are used: the prime indicates the value of a variable after selection, or in the next generation; $\Delta x$ is the change in value of variable $x$ after selection or between generations; the statistical functions $\text{Reg}[x]$, $\text{Var}[x]$, $\text{Cov}[x]$, $\text{E}[x]$ are the regression, variance, covariance, and expected value functions of $x$, respectively (often it is simpler to denote the expected value of $x$ as $\bar{x}$); when necessary to avoid confusion in writing statistical functions, the vector of weights used is given as a bold subscript to the operator (so, for example, in equation 5-2 $\text{Cov}_{\mathbf{q}}[W_i, q_i]$ means the covariance of individual fitness with individual gene frequency using the vector of gene frequencies $\mathbf{q} = (1 - q, q)$ as weights in calculating the covariance); $dx/dy$ refers to the derivative of variable $x$ with respect to $y$, while $\dot{x}$ is understood to be the derivative of variable $x$ with respect to time (if partial derivatives are required, then the notation $\partial F(x,y)/\partial x$ is used); $\hat{x}$ is the equilibrium frequency of variable $x$.

In the following list, terms and variables are given in order of appearance in the book, organized by chapter.

*Chapter 1*

| | |
|---|---|
| $X_i$ | Density or frequency of type $i$ |
| $t$ | Time |

| | |
|---|---|
| $F_i$ | Fisher's fitness, $F$-fitness |
| $m$ | Fisher's Malthusian parameter |
| $b_i, d_i$ | Heritable capacities at birth and avoiding death, respectively, for type $i$ |
| $q_i, q_i'$ | Value of a property of interest for type $i$ |
| $f_i, f_i'$ | Frequency (or some other measure of abundance) for type $i$ in the pre- and post-selection populations |
| $\bar{q}, \bar{q}'$ | Average property in the population before and after selection |
| $W_i$ | Fitness as a ratio of the frequencies before and after selection |

## Chapter 2

| | |
|---|---|
| $X_i, \dot{X}_i$ | Density of molecule of type $i$ and change in density with time |
| $\beta_i, \delta_i$ | Forward and backward rates of the reaction converting substrate to type $i$ |
| $R$ | Concentration of resources in the local environment |
| $b_i, d_i, R_i$ | Heritable capacities at birth, avoiding death, and accruing resources, respectively, for type $i$ |
| $\hat{K}_i$ | Carrying capacity for type $i$ as the density the replicator would attain if it were alone in the environment at equilibrium |
| $R_T$ | Total resources available in the population |

## Chapter 3

| | |
|---|---|
| $X_i, \dot{X}_i$ | Density of replicator of type $i$ and change in density with time |
| $I_i, E_i$ | Informational molecule (replicator $i$) and the enzyme produced by replicator $i$, respectively |
| $B_i, B$ | Beneficial effect of enzymes on a replicator's rate of replication |
| $b_i$ | Rate of template-mediated replication for replicator $i$ |
| $\Psi$ | Per capita production of the community of replicators |
| $d_i$ | Rate of death for replicator $i$ |
| $r_i$ | Defined for convenience, $r_i = b_i - d_i$ |
| $X_T$ | Total density of replicators in the population, $X_T = \Sigma_i X_i$ |
| $\mu$ | Mutation rate in quasispecies |
| $z_2$ | Defined for mathematical convenience, $z_2 = X_2 / X_T$ |

| | |
|---|---|
| $N$ | Population size in local habitats |
| $y, \bar{y}$ | Local density and average density, respectively, of type 2 replicators |
| $C$ | Cost of producing an enzyme catalyst |
| $e_i$ | Experienced density of type 2 replicators for a replicator of type $i$, $i = 1,2$ |
| $\sigma_y^2$ | Variance in local density of type 2 replicators |
| $F_{ST}$ | Wright's statistic, $F_{ST} = \sigma_y^2/\bar{y}(1 - \bar{y})$ |

*Chapter 4*

| | |
|---|---|
| $W_i$ | Individual fitness of genotype $i$ (expected number of gametes produced) |
| $f_i$ | Frequency of genotype $i$ |
| $p, q$ | Frequency of alleles $A$ and a, respectively, in the total population |
| $N$ | Size of total population |
| $\overline{W}$ | Average fitness of the total population |
| $F(q)$ | Wright's "fitness function" |
| $\overline{W}_i$ | Marginal fitness of allele $i$, $i = A,a$ |
| $q_i$ | Frequency of allele $a$ in individual $i$, sometimes referred to as "individual gene frequency" |
| $g_i, n_z$ | Gene dosage in individual $i$ and zygotic ploidy of the species, respectively |
| $\alpha$ | Fisher's average effect of a gene substitution, the regression of fitness on individual gene frequency, $\alpha = \mathrm{Reg}[W_i,q_i]$ |
| $\mu$ | Individual fitness of heterozygote in additive model of fitness in table 4-2 |
| $P, 2Q, R$ | Fisher's notation for frequency of $AA$, $Aa$, and $aa$ genotypes in table 4-2 |
| $W_i^+$ | Individual fitness of genotype $i$ predicted by additive model in table 4-2 |
| $W_{i,s}$ | Fitness of individual $i$ in subpopulation $s$ in table 4-3 |
| $\overline{W}_s$ | Average fitness of subpopulation $s$ in table 4-3 |
| $N_s$ | Number of individuals in subpopulation $s$ in table 4-3 |
| $q_s$ | Frequency of allele in subpopulation $s$ in table 4-3 |
| $k_{ij}$ | Number of cells of type $i$ in the adult form of a zygote of type $j$ in figure 4-1 |
| $k_j, k_j'$ | Total number of cells before and after selection and other processes of change for cell group $j$; in figure 4-1 and chap- |

ters 5 and 6, zygote reproduction is assumed, so $k_j = 1$ and the prime notation is dropped for the adult stage

| | |
|---|---|
| $b$ | Fitness effect at the cell level in table 4-4 |
| $\omega_i$ | Ratios of weights representing transformations in table 4-4 |
| $P(j/i)$ | Conditional probability of interactions defined as the probability of interacting with type $j$ for an individual of type $i$ |
| $w_{ij}$ | Fitness of type $i$ when interacting with type $j$ |
| $c_i, b_i$ | Effect of type $i$'s behavior on the individual fitness of self and others, respectively |
| $c, b$ | Cost (to self) and benefit (to others) of altruistic behavior |
| $R$ | Relatedness in the population |
| $x, \bar{x}$ | Trait value and average trait value |
| $W(x,\bar{x})$ | Fitness of trait $x$ in a population with average trait $\bar{x}$ |
| $R, P, S, T$ | Fitness of interactions in prisoner's dilemma game |
| $\mu_{TFT}, \mu_{AD}$ | Individual mobilities of $TFT$ and $AD$ strategies, respectively |

*Chapter 5*

| | |
|---|---|
| $k_{ij}$ | Number of $i$ cells in the adult stage of a $j$-zygote: $i,j = C,D$ |
| $k_j$ | Total number of cells in the adult stage of a $j$-zygote after development: $j = C,D$ |
| $W_D$ | Absolute individual fitness of a $D$-zygote: $\propto k_D$ |
| $W_C$ | Absolute individual fitness of a $C$-zygote: $\propto k_{CC} + k_{DC} + \beta k_{CC}$ |
| $\beta$ | Benefit to adult organism of cooperation among its cells |
| $q_j, \Delta q_j$ | Initial frequency, and change in frequency, of $C$ gene within $j$ organisms |
| $q, \Delta q$ | Initial frequency, and change in frequency, of $C$ gene in total population |
| $h_w^2$ | Heritability of fitness at the cell-group or organism level; see table 5-2 |
| $\mu$ | Within-organism mutation rate from $C$ to $D$ per cell division |
| $t$ | Time for development |
| $c$ | Rate of cell division for cooperating cells ($c = 1$ in most analyses) |
| $b$ | Advantage to cell of defection (in terms of the cell's replication rate). Although the results discussed in the book primarily relate to mutations that are disadvantageous at the organism level and advantageous at the cell level ($b > 1$), the model may consider any nonnegative values of $b$. When $b$ |

= 1, the mutations are neutral at the cell level; when $0 < b < 1$, the mutations are deleterious at the cell level; and when $b = 0$, the mutations do not replicate but may remain in the organism. Mutations that are deleterious or neutral at the cell level also select for a germ line and other kinds of conflict mediation (see Michod and Roze 1998).

| | |
|---|---|
| $cb$ | Rate of cell division for defecting cells |
| $s_C, s_D$ | Probability of cell death for $C$ and $D$ cells, respectively |
| $\lambda$ | Eigenvalues |
| $w, \hat{w}$ | Relative individual fitness and average relative fitness at equilibrium plotted in figure 5-4; $\hat{w} = W_C h_W^2 / W_D$ |

*Chapter 6 (notation in addition to that defined in chapter 5)*

| | |
|---|---|
| $i, j$ | Index for genotype 1, 2, 3, 4 = $CM, Cm, DM, Dm$ |
| $k_{ij}$ | Number of $i$ cells in the adult stage (soma) of a $j$-zygote |
| $k_j$ | Total number of cells in the adult stage (soma) of a $j$-zygote |
| $K_{ij}$ | Number of $i$ cells in the germ line of a $j$-zygote |
| $K_j$ | Total number of cells in the germ line of a $j$-zygote |
| $W_j$ | Individual fitness of a $j$-zygote: $W_j = k_j + \beta(k_{1j} + k_{2j})$ |
| $r$ | Recombination rate between $C/D$ and $M/m$ loci |
| $G$ | Linkage disequilibrium defined in equation 6-1 |
| $x_j$ | Frequency of a two-locus $j$ genotype in total population |
| $\delta$ | Effect of modifier: reduction in development time in germ line (germ line modifier) or cost of policing on organism fitness (self-policing modifier) |
| $\varepsilon$ | Effect of policing modifier on cell fitness |
| $\mu, \mu_M$ | Within-organism mutation rate in soma and germ line, respectively (germ line modifier) |
| $t, t_M$ | Time for development in the soma and germ line, respectively (germ line modifier) |
| $W_P, W_O$ | Individual fitness of parents and offspring, respectively |

*Chapter 7*

| | |
|---|---|
| $s, k$ | Survival and fecundity in cost of sex model |
| $P, R$ | Frequency of sexual and asexual alleles |
| $P, 2Q, R$ | Frequency of $AA, Aa,$ and $aa$ genotypes in table 7-2 |
| $p$ | Frequency of $A$ allele defined in table 7-2 |
| $W_{ij,O}$ | Individual fitness of genotype $ij$ among offspring |

| | |
|---|---|
| $P(j/i)$ | Conditional probability of interactions defined as the probability of interacting with type $j$ for an individual of type $i$ |
| $c_i, b_i$ | Effect of type $i$'s behavior on the fitness of self and others, respectively |
| $X_1, X_2, X_3$ | Frequencies of genotypes $AA$, $Aa$, and $aa$, respectively, in figure 7-1 |
| $W_1, W_2, W_3$ | Individual fitnesses of genotypes $AA$, $Aa$, and $aa$, respectively, in figure 7-1 |
| $F(y)$ | Fisherian fitness of trait $y$ in figure 7-2 |
| $y_0, s$ | Specific values of trait $y$ in figure 7-2: $y = s$ is the optimum, $y = y_0$ is the current state |

### Chapter 8

| | |
|---|---|
| $X_i, \dot{X}_i$ | Density of type $i$ and change in density with time |
| $\Psi$ | Per capita production of the community of replicators |
| $b_i, d_i$ | Heritable capacities at birth and avoiding death, respectively, for type $i$ |
| $W_i$ | Individual fitness of type $i$ |
| $\beta$ | Benefit to adult organism of cooperation among cells (in reference to results in chapters 5 and 6) |
| $t$ | Time for development (in reference to results in chapters 5 and 6) |
| $b$ | Advantage to cell of defection (in reference to results in chapters 5 and 6) |
| $R$ | Level of resources in the local environment |
| $F_i(\mathbf{a}_i, \mathbf{E})$ | Fisherian fitness of type $i$ as a function of heritable capacities and the environment |

### Appendix A (notation in addition to that defined in chapters 5 and 6)

| | |
|---|---|
| $i, j$ | Indices for diploid genotypes for cells and zygotes: $i, j = 0, 1, 2$ for $DD$, $CD$, $CC$ |
| $f_i$ | Frequency of genotype $i$ in the population |
| $W_i$ | Individual fitness of genotype $i$ |
| $d, h$ | Dominance parameters at level of adult organism and cell, respectively |
| $\overline{W}$ | Average individual fitness: $\overline{W} = \Sigma_i f_i W_i$ |
| $\mathbf{f}^i$ | Vectors of initial frequencies of the three genotypes in each of the three zygotes: $i = DD, CD, CC$. Used in study of fitness statistics for asexual diploids |

231

$P, 2Q, R$        Frequency of $gg$, $Gg$, and $GG$ genotypes in tables A-5 and A-6

$h$        Dominance parameter, probability that $Gg$ heterozygotes produce offspring sexually

$p_i, o_i$        Frequencies of allele $i$ in sperm and eggs, respectively: $i = G, g$

# Notes

PREFACE

1. See Dawkins (1982). This is further discussed in chapter 7, "Reconsidering Fitness."

CHAPTER 1
THE LANGUAGE OF SELECTION

1. For discussion of Darwin's conditions as a basis for natural selection, see Endler 1986, chap. 1.

2. Individual fitness goes by various names, such as *genotypic fitness, Wrightian fitness,* and *Darwinian fitness.* Sometimes the absolute fitnesses of the different genotypes ("absolute" because absolute numbers of gametes are counted) are expressed relative to a standard type (say, the *AA* homozygote in the case of selection at a single gene locus with two alleles *A* and *a*) by dividing through by the absolute Darwinian fitness of the *AA* homozygote, $W_{AA}$. Typically, the resulting *relative fitnesses* are expressed as deviations from the standard, as 1, $1 - s_{Aa}$, and $1 - s_{aa}$, for *AA, Aa,* and *aa.* The deviations are referred to as the *selective values* of the genotypes. Although commonly used in standard population genetics, relative fitness and selective values are of little use for the study of selection in hierarchically structured populations. This is because relative fitness tends to obscure the contribution of the group to the total change in the global population. In addition, absolute fitness provides the connection between genetical selection and population growth in ecology that I explore in chapter 4 (see also Ginzburg 1983). In figures 5-4 and 5-5 relative fitness is used so as to compare fitness statistics at the emerging organism level.

CHAPTER 2
ORIGIN OF FITNESS

1. This matter is discussed in more detail in chapter 7, "The 'Tautology' Problem."

CHAPTER 3
THE FIRST INDIVIDUALS

1. It is worth noting that it is not necessary to assume the existence of a complex translation apparatus for a replicator to make a primitive protein, as is sometimes claimed (Barbieri 1981). The first protein may have been produced by direct physical-chemical interactions between the nucleotide bases and amino acids (Hendry et al. 1981a,b, Woese 1967).

2. This is discussed in more detail in chapter 4, "The Prisoner's Dilemma."

3. I wish to acknowledge the role of Fred Hopf in setting up this model. We were working on it together in 1981 but his premature death prevented us from completing its analysis together.

CHAPTER 4
EVOLUTION OF INTERACTIONS

1. There are other problems with the third phase of the shifting-balance process, even when selection is independent of local gene frequencies (Coyne et al. 1997; Gavrilets 1996; Haldane 1959).

2. In Fisher (1941), the regression is half that calculated here, since Fisher uses allele dosage (0, 1, 2) instead of allele frequency (1, 1/2, 1). The additive effect on fitness is then $\alpha$ instead of $\alpha/2$ as in table 4-2. When there are multiple alleles at multiple loci, the relationship between Fisher's average effect and the regression coefficient is more complicated (see Price 1972b, eq. 5.3).

3. See also Fisher 1941, p. 56, but note the typo in the middle of the page where $P + 2Q + R$ in the denominator should be raised to the power 2. I have expressed Fisher's theorem in terms of the average effect and the variance in individual gene frequency, instead of in terms of the average effect and the average excess, as Fisher usually did (see, for example, Fisher 1941, eq. 5).

4. The view that natural selection should increase individual fitness is based on the commonsense notion of selection.

5. I discuss the implications of this fact for Wright's shifting-balance process in the section "Frequency Dependence Decouples Fitness in a Selection Hierarchy" earlier in this chapter.

CHAPTER 5
MULTILEVEL SELECTION OF THE ORGANISM

1. Although $\beta$ does not affect the change within an organism, it does affect the fitness of $C$ adults and, consequently, the average taken in equation 5-2. So there is an effect of $\beta$ in panel ($C$) of figure 5-5 when the limiting value is reached.

2. See, for example, Michod 1995, p. 119, fig. 1 and the associated discussion.

3. This will change, however, when I consider two-locus haploid modifier models in the next chapter.

4. Figures 8–12 of Michod (1997a) repeat the analysis of figures 5-3–5-5 for diploidy.

CHAPTER 7
FITNESS EXPLANATIONS

1. Certain far-from-equilibrium physical phenomena such as lasers are an exception (Bernstein et al. 1983).

2. By asexual reproduction, I mean cloning in which parents produce offspring of identical genotypes. I do not consider here the many forms of meiotic parthenogenesis, which, although important for understanding the evolution of sex, are not central to my main focus concerning fitness and its role in evolutionary explanations of sexual and asexual reproduction.

3. In the present discussion there is just one level of selection, the organism. As we discovered in chapter 5 (see also the appendix), the effect of sex on the heritability of fitness is vastly more complicated when mutation and selection are occurring at several levels simultaneously. Sex can help maintain high heritability of fitness at the organism level when there is a lot of within-organism change.

4. Used in this way, the term "germ line" refers to the cell lineage that can be traced backward in time from any living thing, whether or not it is a multicellular organism and possesses a germ line, in the sense of a sequestered and differentiated cell type specialized in forming gametes, and a somatic line with terminal differentiation.

5. Other details are being ignored for reasons of space. Michod (1982) provides a review of these details.

6. This notion of fitness was termed "personal fitness" by Orlove (1975; Orlove and Wood 1978) and "neighbour-modulated fitness" by Hamilton (1964a).

7. For an introduction to family-structured models, see, for example, Michod (1982).

8. Here I follow an earlier presentation of mine given in Michod (1995, chap. 9 and appendix).

CHAPTER 8
A PHILOSOPHY OF FITNESS

1. Endler also notes the fundamental distinction between success- and design-based definitions of fitness (Endler 1986, p. 38).

2. Specifically, equations 2-2, 3-2, 3-4, 3-5, 4-1, 4-5, 4-16, 5-1, 6-1, 6-2, and 8-1 below.

3. A similar example to Eigen's equation 3-2 is considered below in equation 8-1 as a means to introduce the general schema proposed in equation 8-2.

4. In previous work (Byerly and Michod 1991a; Michod 1986, 1984) the heritable capacities were termed "adaptive" or "intrinsic" capacities, but as discussed in the section "$F$-Fitness and Evolutionary Explanations" below, I think this terminology is confusing and misleading.

5. See also the discussion of supervenience in the section "Heritable Capacities as Components of Design" below.

6. The prisoner's dilemma game is discussed in chapter 4.

7. I briefly discuss long-term measures of fitness above.

8. Maynard Smith agreed with the main conclusion of that paper, which was that overall adaptedness of organisms does not enter into the theory of evolution.

9. Recall equations 5-1, 5-3, and 5-4, and see tables A-3 and A-4.

10. As discussed chapter 4, Fisher's fundamental theorem embodies the desire to separate the effects of "selection" from the effects of changes in the genetic, biotic, and physical environment. Price notes that Fisher's derivation of his fundamental theorem requires the absence of lower-level change (Price 1972b).

11. This dynamic is similar to equations 1-2 and 2-2 except in the way in which resources, $R$, are incorporated. Resources are assumed to be a decreasing function of the total density of types, for example, $R = 1 - \Sigma X_i$.

12. Much of Brandon's discussion concerns the meaning of "intrinsic." In previous work I referred to heritable capacities as intrinsic (Michod 1986, 1984), but in later work dropped the term entirely (Byerly and Michod 1991a). Nevertheless, the criticisms Brandon offers apply to heritability, since heritability may also change with the environment. These criticisms are addressed in the text.

13. Admittedly we would want a two-sex model of selection, not equation 8-1, but I hope my point is clear.

14. For example, the right-hand side of equations 2-2, 3-2, 3-4, 3-5, 5-1, 6-1, 6-2, or 8-1.

15. To see that this is not generally possible, consider the underlying mathematical problem. Suppose $F$ is a function of three variables: $F(a_1, a_2, e)$. It is not usually possible to construct another function $F = G[A(a_1, a_2), e]$, in which $A$ (for adaptedness) is a function of only the heritable capacities $a_1$ and $a_2$.

16. Heritability, too, may change with the environment (see the discussion above in "Brandon's Approach").

# References

Abrams, P. A., Matsuda, H., and Harada, Y. 1993. Evolutionarily unstable fitness maxima and stable fitness minima of continuous traits. *Evolutionary Ecology* 7:465–487.

Akopyants, N. S., Eaton, K. A., and Berg, D. E. 1995. Adaptive mutation and cocolonization during *Helicobacter pylori* infection of gnotobiotic piglets. *Infectious Immunology* 63:116–121.

Albert, B., Godelle, B., Atlan, A., De Paepe, R., and Gouyon, P. H. 1996. Dynamics of plant mitochondrial genome: Model of a three-level selection process. *Genetics* 144:369–382.

Ameisen, J. C. 1996. The origin of programmed cell death. *Science* 272:1278–1279.

American Genetics Society Symposium for the Evolution of Sex. 1993. *Journal of Heredity* 84.

Anderson, P. 1997. Kinase cascades regulating entry into apoptosis. *Microbiological Reviews* 61:33–46.

Anderson, W. W. 1969. Polymorphism resulting from the mating advantage of rare male genotypes. *Proceedings of the National Academy of Sciences* 64:190–197.

———. 1971. Genetic equilibrium and population growth under density-regulated selection. *American Naturalist* 105:489–498.

Anderson, W. W. and King, C. E. 1970. Age-specific selection. *Proceedings of the National Academy of Sciences* 66:780–786.

Asmussen, M. A. 1979. Density dependent selection 2. The Allee effect. *American Naturalist* 114:796–809.

———. 1983a. Density-dependent selection incorporating intraspecific competition 2. A diploid model. *Genetics* 103:335–350.

———. 1983b. Density dependent selection incorporating intraspecific competition 1. a haploid model. *Journal of Theoretical Biology* 101:113–127.

Asmussen, M. A. and Feldman, M. W. 1977. Density dependent natural selection 1: A stable feasible equilibrium may not be attainable. *Journal of Theoretical Biology* 64:603–618.

Axelrod, R. 1984. *Evolution of Cooperation.* New York: Basic Books.

Axelrod, R. and Hamilton, W. D. 1981. The evolution of cooperation. *Science* 211:1390–1396.

Barbieri, M. 1981. The ribotype theory on the origin of life. *Journal of Theoretical Biology* 91:545–601.

Barton, N. H. 1992. On the spread of a new gene combination in the third phase of Wright's shifting-balance. *Evolution* 45:499–517.

Belhassen, E., Dommee, B., Atlan, A., Gouyon, P. H., Pomente, D., Assouad, M. W., and Couvet, D. 1991. Complex determination of male sterility in *thymus-vulgaris* I. Genetic and molecular analysis. *Theoretical and Applied Genetics* 82:137–143.

237

Bell, G. 1985. The origin and early evolution of germ cells as illustrated by the Volvocales. Pages 221–256 *The Origin and Evolution of Sex,* edited by H. O. Halvorson and A. Monroy. New York: Alan R. Liss, Inc.

Bell, G. and Koufopanou, V. 1991. The architecture of the life cycle in small organisms. *Philosophical Transactions of the Royal Society of London B, Biological Sciences* 332:81–89.

Bernstein, C. and Bernstein, H. 1991. *Aging, Sex and DNA Repair.* San Diego: Academic Press.

Bernstein, H., Byerly, H. C., Hopf, F. A., and Michod, R. E. 1984. Origin of sex. *Journal of Theoretical Biology* 110:323–351.

———. 1985a. DNA damage, mutation and the evolution of sex. *Science* 229:1277–1281.

———. 1985b. The evolutionary role of recombinational repair and sex. *International Review of Cytology* 96:1–28.

———. 1985c. Sex and the emergence of species. *Journal of Theoretical Biology* 117:665–690.

Bernstein, H., Byerly, H. C., Hopf, F. A., Michod, R. E., and Vemulapalli, G. K. 1983. The Darwinian dynamic. *Quarterly Review of Biology* 58:185–207.

Bernstein, H., Hopf, F. A., and Michod, R. E. 1987. The molecular basis of the evolution of sex. *Advances in Genetics* 24:323–370.

Blackstone, N. W. and Ellison, A. M. 1998. Metazoan development and levels of selection. Unpublished manuscript.

Blattner, F. R., Bloch, C. A., Perna, N. T., Burland, V., Riley, M., Collado-Vides, J., Glasner, J. D., Rode, C. K., Mayhew, G. F., et al. 1997. The complete genome sequence of *Escherichia coli* K-12. *Science* 277 (5331):1453–1744.

Blumenthal, E. Z. 1992. Could cancer be a physiological phenomenon rather than a pathological misfortune? *Medical Hypotheses* 39:41–48.

Boerlijst, M. C. and Hogeweg, P. 1991. Spiral wave structure in prebiotic evolution: Hypercycles stable against parasites. *Physica D* 48:17–28.

———. 1993. Evolutionary consequences of spiral waves in a host-parasitoid system. *Proceedings of the Royal Society of London B, Biological Sciences* 253:15–18.

Boyd, R. and Lorberbaum, J. 1987. No pure strategy is evolutionarily stable in the repeated Prisoner's Dilemma game. *Nature (London)* 327:58–59.

Boyd, R. and Richerson, P. J. 1992. Punishment allows the evolution of cooperation (or anything else) in sizable groups. *Ethology and Sociobiology* 13:171–195.

Brandon, R. N. 1978. Adaptation and evolutionary theory. *Studies in the History and Philosophy of Sciences* 9:188–206.

———. 1980. A structural description of evolutionary theory. Pages 427–439 in *PSA 1980,* Vol. 2, edited by P. Asquith and R. Giere. East Lansing: Philosophy of Science Association.

———. 1991. *Adaptation and Environment.* Princeton, N.J.: Princeton University Press.

Brown, J. S., Sanderson, M. J., and Michod, R. E. 1982. Evolution of social behavior by reciprocation. *Journal of Theoretical Biology* 99:319–339.

Burns, T. P. 1992. Adaptedness, evolution and a hierarchical concept of fitness. *Journal of Theoretical Biology* 154:219–238.

Buss, L. W. 1987. *The Evolution of Individuality.* Princeton, N.J.: Princeton University Press.

Byerly, H. C. and Michod, R. E. 1991a. Fitness and evolutionary explanation. *Biology and Philosophy* 6:1–22.

———. 1991b. Fitness and evolutionary explanation: a response. *Biology and Philosophy* 6:45–53.

Carson, D. A. and Ribeiro, J. M. 1993. Apoptosis and disease. *Lancet* 341:1251–1254.

Cech, T. R. 1986. A model for the RNA-catalyzed replication of RNA. *Proceedings of the National Academy of Sciences* 83:4360–4363.

———. 1987. The chemistry of self-splicing RNA and RNA enzymes. *Science* 236:1532–1539.

Charlesworth, B. 1980. *Evolution in Age-Structured Populations.* Cambridge, England: Cambridge University Press.

Chigira, M. and Watanabe, H. 1994. Differentiation as symbiosis. *Medical Hypotheses* 43:17–18.

Clarke, G. M. and McKenzie, J. A. 1987. Developmental stability of insecticide resistant phenotypes in blowfly; a result of canalizing natural selection. *Nature (London)* 325:345–346.

Cooke, F., Taylor, P. D., Francis, C. M., and Rockwell, R. F. 1990. Directional selection and clutch size in birds. *American Naturalist* 136:261–267.

Cooper, W. S. 1984. Expected time to extinction and the concept of fundamental fitness. *Journal of Theoretical Biology* 107:603–629.

Coppes, M. J., Liefers, G. J., Paul, P., Yeger, H., and Williams, B. R. G. 1993. Homozygous somatic WT1 point mutations in sporadic unilateral Wilms' tumor. *Proceedings of the National Academy of Sciences* 90:1416–1419.

Cox, M. M. 1991. The RecA protein as a recombinational repair system. *Molecular Microbiology* 5:1295–1299.

Coyne, J. A., Barton, N. H., and Turelli, M. 1997. Perspective: A critique of Sewall Wright's shifting balance theory of evolution. *Evolution* 51:643–671.

Crow, J. F. 1970. Genetic loads and the cost of natural selection. Pages 128–177 in *Mathematical Topics in Population Genetics,* edited by K. Kojima. Berlin: Springer-Verlag.

Crow, J. F., Engels, W. R., and Denniston, C. 1990. Phase three of Wright's shifting-balance. *Evolution* 44:233–247.

Crow, J. F. and Kimura, M. 1970. *Introduction to Population Genetics Theory.* Minneapolis, Minn.: Burgess.

Darwin, Charles. 1859. *On the Origin of Species by Means of Natural Selection, or the Preservation of Favoured Races in the Struggle for Life.* 1st ed. London: John Murray.

239

REFERENCES

————. 1872. *The Origin of Species by Means of Natural Selection, or the Preservation of Favoured Races in the Struggle for Life.* 6th ed. London: John Murray.

Davidson, E. H., Peterson, K. J., and Cameron, R. A. 1995. Origin of bilateral body plans: evolution of developmental regulatory mechanisms. *Science* 270:1319–1325.

Dawkins, R. 1982. *Extended Phenotype: The Gene as the Unit of Selection.* Oxford: W. H. Freeman.

De Feo, O. and Ferriere, R. 1997. Bifurcation analysis of invasion in a simple competitive model. Manuscript submitted to *Journal of Mathematical Biology.*

Demerec, M. 1936. Frequency of "cell-lethals" among lethals obtained at random in the X-chromosome of *Drosophila melanogaster. Proceedings of the National Academy of Sciences* 22:350–354.

Dennis, J., Donaghue, T., Florian, M., and Kerbel, R. S. 1981. Apparent reversion of stable in vitro genetic markers detected in tumour cells from spontaneous metastases. *Nature (London)* 292:242–245.

Denniston, C. 1978. An incorrect definition of fitness revisited. *Annals of Human Genetics* 42:77–85.

Dewey, J. 1909. Darwin's influence upon philosophy. *The Popular Science Monthly* July:90–98.

Diekmann, O., Mylius, S. D., and ten Donkelaar, J. R. 1998. Saumon à la Kaitala et Getz, sauce hollandaise. *Evolutionary Ecology* (in press).

Dobzhansky, T. 1969. On Cartesian and Darwinian aspects of biology. Pages 165–178 in *Philosophy, Science, and Method,* edited by S. Morganbesser, P. Suppes, and M. White. New York: St. Martin's Press.

Drake, J. W. 1991. A constant rate of spontaneous mutation in DNA-based microbes. *Proceedings of the National Academy of Sciences* 88:7160–7164.

Dugatkin, L. A. and Wilson, D. S. 1991. ROVER: A strategy for exploiting cooperators in a patchy environment. *American Naturalist* 138:687–701.

Du Mouchel, W. H. and Anderson, W. W. 1968. The analysis of selection in experimental populations. *Genetics* 58:435–449.

Edelman, G. M. 1987. *Neural Darwinism.* New York: Basic Books.

Edwards, A. W. F. 1990. Fisher, Wbar, and the fundamental theorem. *Theoretical Population Biology* 38:276–284.

Eigen, M. 1971. Self-organization of matter and the evolution of biological macromolecules. *Naturwissenschaften* 58:465–523.

————. 1992. *Steps Towards Life.* Oxford: Oxford University Press.

Eigen, M., Gardiner, W., Schuster, P., and Winkler-Oswatitsh, R. 1981. The origin of genetic information. *Scientific American* 244 (4):88–118.

Eigen, M. and Schuster, P. 1977. The hypercycle, a principle of natural self-organization. Part A: emergence of the hypercycle. *Naturwissenschaften* 64:541–565.

————. 1978a. The hypercycle, a principle of natural self-organization. Part B: the abstract hypercycle. *Naturwissenschaften* 65:7–41.

————. 1978b. The hypercycle, a principle of natural self-organization. Part C: the realistic hypercycle. *Naturwissenschaften* 65:341–369.

240

————. 1979. *The Hypercycle, a Principle of Natural Self-Organization.* Berlin: Springer-Verlag.

Endler, J. A. 1986. *Natural Selection in the Wild.* Princeton, N.J.: Princeton University Press.

Enquist, M. and Leimar, O. 1993. The evolution of cooperation in mobile organisms. *Animal Behavior* 45:747–757.

Enserink, M. 1997. Thanks to a parasite, asexual reproduction catches on. *Science* 275:1743.

Eshel, I. 1996. On the changing concept of evolutionary population stability as a reflection of a changing point of view in the quantitative theory of evolution. *Journal of Mathematical Biology* 34:485–510.

Eshel, I. and Cavalli-Sforza, L. L. 1982. Assortment of encounters and the evolution of cooperativeness. *Proceedings of the National Academy of Sciences* 79:1331–1335.

Fagerstrom, T. 1992. The meristem-meristem cycle as a basis for defining fitness in clonal plants. *Oikos* 63:449–453.

Falconer, D. S. 1989. *Introduction to Quantitative Genetics.* 3d ed. Burnt Mill, England: Longman.

Farber, E. 1984. Cellular biochemistry of the stepwise development of cancer with chemicals: G. H. A. Clowes memorial lecture. *Cancer Research* 44:5463–5474.

Ferriere, R. and De Feo, O. 1998. Invasion dynamics and the definition of fitness in populations with multiple invariant states. Unpublished manuscript.

Ferriere, R. and Michod, R. E. 1995. Invading wave of cooperation in spatial iterated prisoner's dilemma. *Proceedings of the Royal Society of London B, Biological Sciences* 259:77–83.

————. 1996. The evolution of cooperation in spatially heterogeneous populations. *American Naturalist* 147:692–717.

————. 1998. The analysis of wave patterns predicts cooperation in spatially distributed populations. In *Advances in Adaptive Dynamics,* edited by U. Dieckmann and J. A. J. Metz. Cambridge, England: Cambridge University Press (in press).

Fisher, R. A. 1930. *The Genetical Theory of Natural Selection.* Oxford: The Clarendon Press.

————. 1941. Average excess and average effect of a gene substitution. *Annals of Eugenics* 11:53–63.

————. 1958. *The Genetical Theory of Natural Selection.* New York: Dover.

Fox, S. W. and Harada, K. 1960. Thermal copolymerization of amino acids common to proteins. *Journal of the American Chemical Society* 82:3745–3751.

Frank, S. A. 1995. George Price's contributions to evolutionary genetics. *Journal of Theoretical Biology* 175:373–388.

Frank, S. A. and Slatkin, M. 1992. Fisher's fundamental theorem of natural selection. *Trends in Ecology and Evolution* 7:92–95.

Gatenby, R. A. 1991. Population ecology issues in tumor growth. *Cancer Research* 51:2542–2547.

Gaul, H. 1958. Present aspects of induced mutations in plant breeding. *Euphytica* 7:275–289.

Gavrilets, S. 1996. On phase three of the shifting-balance theory. *Evolution* 50: 1034–1041.

Gayley, T. and Michod, R. E. 1990. Modification of genetic constraints on frequency-dependent selection. *American Naturalist* 136:406–427.

Geritz, S. A. H., Kisdi, É., Meszéna, G., and Metz, J. A. J. 1998. Evolutionarily singular strategies and the adaptive growth and branching of the evolutionary tree. *Evolutionary Ecology* 12:35–57.

Geritz, S. A. H., Metz, J. A. J., Kisdi, É., and Meszéna, G. 1997. The dynamics of adaptation and evolutionary branching. *Physical Review Letters* 78:2024–2027.

Gillespie, J. H. 1973. Natural selection with varying selection coefficients—a haploid model. *Genetical Research* 21:115–120.

———. 1974. Natural selection for within-generation variance in offspring. *Genetics* 76:601–606.

———. 1977. Natural selection for variances in offspring numbers—a new evolutionary principle. *American Scientist* 111:1010–1014.

Ginzburg, L. R. 1983. *Theory of Natural Selection and Population Growth.* Menlo Park, Calif.: Benjamin/Cummings.

Glansdorff, P. and Prigogine, I. 1971. *Thermodynamic Theory of Structure, Stability and Fluctuations.* London: J. Wiley and Sons.

Gliddon, C. J. and Gouyon, P. H. 1989. The units of selection. *Trends in Ecology and Evolution* 4:204–208.

Godelle, B. and Reboud, X. 1995. Why are organelles uniparentally inherited? *Proceedings of the Royal Society of London B, Biological Sciences* 259:27–33.

Gouyon, P. H. and Couvet, D. 1985. Selfish cytoplasm and adaptation: Variations in the reproductive system of thyme. Chapter 20, pp. 299–319 in *Structure and Functioning of Plant Populations II,* edited by J. W. Woldendorp and J. Haeck. Amsterdam: North-Holland.

Gouyon, P. H., Vichot, F., and Van Damme, J. 1991. Nuclear-cytoplasmic male sterility single-point equilibria versus limit cycles. *American Naturalist* 137:498–514.

Hague, A., Manning, A. M., van der Stappen, J. W., and Paraskeva, C. 1993. Escape from negative regulation of growth by transforming growth factor beta and from the induction of apoptosis by the dietary agent sodium butyrate may be important in colorectal carcinogenesis. *Cancer Metastasis Review* 12:227–237.

Haldane, J. B. S. 1929. The origin of life. *Rationalist Annual* 1929:148–169.

———. 1937. The effect of variation on fitness. *American Naturalist* 71:337–349.

———. 1959. Natural selection. Pages 101–149 in *Darwin's Biological Work: Some Aspects Reconsidered,* edited by P. R. Bell. New York: Wiley.

Haldane, J. B. S. and Jayakar, S. D. 1963. Polymorphism due to selection of varying direction. *Journal of Genetics* 58:237–242.

Hamilton, W. D. 1963. The evolution of altruistic behavior. *American Naturalist* 97:354–356.

————. 1964a. The genetical evolution of social behaviour. I. *Journal of Theoretical Biology* 7:1–16.

————. 1964b. The genetical evolution of social behaviour. II. *Journal of Theoretical Biology* 7:17–52.

————. 1975. Innate social aptitudes of man: an approach from evolutionary genetics. Pages 133–155 in *Biosocial Anthropology*, edited by R. Fox. New York: Wiley.

Hammerstein, P. 1997. Darwinian adaptation, population genetics, and the streetcar theory of evolution. *Journal of Mathematical Biology* 34:511–532.

Hardin, G. 1968. The tragedy of the commons. *Science* 162:1243–1248.

Hassell, M. P., Comins, H. N., and May, R. M. 1991. Spatial structure and chaos in insect population dynamics. *Nature (London)* 353:255–258.

Hastings, I. M. 1992. Population genetic aspects of deleterious cytoplasmic genomes and their effect on the evolution of sexual reproduction. *Genetical Research* 59:215–225.

Hendry, L. B., Bransome, E. D., Jr., Hutson, M. S., and Campbell, L. K. 1981a. First approximation of a stereochemical rationale for the genetic code based on the topography and physicochemical properties of "cavities" constructed from models of DNA. *Proceedings of the National Academy of Sciences* 78:7440–7444.

Hendry, L. B., Bransome, E. D., Jr., and Petersheim, M. 1981b. Are there structural analogies between amino acids and nucleic acids? *Origins of Life* 11:203–221.

Hoekstra, R. F. 1990. Evolution of uniparental inheritance of cytoplasmic DNA. Pages 269–278 in *Organizational Constraints on the Dynamics of Evolution*, edited by J. Maynard Smith and G. Vida. New York: Manchester University Press.

Hofbauer, J., Schuster, P., and Sigmund, K. 1979. A note on evolutionary stable strategies and game dynamics. *Journal of Theoretical Biology* 81:609–612.

Hoff-Olsen, P., Meling, G. I., and Olaisen, B. 1995. Somatic mutations in VNTR-locus D1S7 in human colorectal carcinomas are associated with microsatellite instability. *Human Mutation* 5:329–332.

Holland, J. H. 1992. *Adaptation in Natural and Artificial Systems: An Introductory Analysis with Applications to Biology, Control, and Artificial Intelligence*. Cambridge, Mass.: MIT Press.

Hopf, F. A. and Hopf, F. W. 1985. The role of the Allee effect in species packing. *Theoretical Population Biology* 27:27–50.

Hopf, F., Michod, R. E., and Sanderson, M. J. 1988. The effect of reproductive system on mutation load. *Theoretical Population Biology* 33:243–265.

Houston, A. I. 1993. Mobility limits cooperation. *Trends in Ecology & Evolution* 8:194–196.

Hughes, R. H. 1989. *A Functional Biology of Clonal Animals*. London: Chapman and Hall.

Hull, D. L. 1981. Individuality and selection. *Annual Review of Ecology and Systematics* 11:311–332.

Hurst, L. D. 1990. Parasitic diversity and evolution of diploidy, multicellularity and isogamy. *Journal of Theoretical Biology* 144:429–443.

REFERENCES

Ionov, Y., Peinado, M. A., Malkhosyan, S., Shibata, D., and Perucho, M. 1993. Ubiquitous somatic mutations in simple repeated sequences reveal a new mechanism for colonic carcinogenesis. *Nature* 363:558–561.

Jablonka, E. and Lamb, M. J. 1995. *Epigenetic Inheritance and Evolution. The Lamarckian Dimension.* Oxford: Oxford University Press.

Jablonski, D. 1986. Larval ecology and macroevolution in marine invertebrates. *Bulletin of Marine Science* 39:565–587.

———. 1997. Extinctions in the fossil record. Pages 25–44 in *Extinction Rates,* edited by J. H. Lawton and R. H. May. Oxford: Oxford University Press.

Kauffman, S. A. 1993. *The Origins of Order, Self-Organization and Selection in Evolution.* New York: Oxford University Press.

Klekowski, E. J., Jr. and Kazarinova-Fukshansky, N. 1984. Shoot apical meristems and mutation: selective loss of disadvantageous cell genotypes. *American Journal of Botany* 71:28–34.

Kondrashov, A. S. 1988. Deleterious mutations and the evolution of sexual reproduction. *Nature (London)* 336:435–440.

———. 1992. The third phase of Wright's shifting balance: a simple analysis of the extreme case. *Evolution* 46:1972–1975.

Kuhn, H. 1972. Self-organization of molecular systems and evolution of the genetic apparatus. *Angewandte Chemie, International Edition in English* 11:798–820.

Kuick, R. D., Neel, J. V., Strahler, J. R., Chu, E. H. Y., Bargal, R., Fox, D. A., and Hanash, S. M. 1992. Similarity of spontaneous germinal and *in vitro* somatic cell mutation rates in humans: Implications for carcinogenesis and for the role of exogenous factors in "spontaneous" germinal mutagenesis. *Proceedings of the National Academy of Sciences* 89:7036–7040.

Kupryjanczyk, J., Thor, A. D., Beauchamp, R., Merritt, V., Edgerton, S. M., Bell, D. A., and Yandell, D. W. 1993. P53 gene mutations and protein accumulation in human ovarian cancer. *Proceedings of the National Academy of Sciences* 90:4961–4965.

Kutter, E., Stidham, T., Guttman, B., Batts, D., Peterson, S., Djavakhishvili, T., Arisaka, F., Mesyanzhinov, V., Ruger, W., and Mosig, G. 1994. Genomic map of bacteriophage T4. Pages 491–519 in *The Molecular Biology of Bacteriophage T4,* edited by J. D. Karam and J. W. Drake. Washington, D.C.: American Society for Microbiology.

Lennox, J. G. 1991. Commentary on Byerly and Michod. *Biology and Philosophy* 6:33–37.

Leon, J. A. 1974. Selection in contexts of interspecific competition. *American Naturalist* 108:739–757.

Leon, J. A. and Charlesworth, B. 1978. Ecological versions of Fisher's fundamental theorem of natural selection. *Ecology* 59:457–464.

Lerner, I. M. 1954. *Genetic Homeostasis.* New York: Dover.

Levene, H. 1953. Genetic equilibrium when more than one ecological niche is available. *American Naturalist* 87:331–333.

Lewontin, R. C. 1970. The units of selection. *Annual Review of Ecology and Systematics* 1:1–18.

244

———. 1972. Testing the theory of natural selection. *Nature (London)* 236:181–182.

———. 1974. *The Genetic Basis of Evolutionary Change.* New York: Columbia University Press.

———. 1978. Adaptation. *Scientific American* 239:212–230.

Li, C. C. 1967a. Fundamental theorem of natural selection. *Nature (London)* 214: 505–506.

———. 1967b. Genetic equilibrium under selection. *Biometrics* 23:367–484.

———. 1976. *First Course in Population Genetics.* Pacific Grove, Calif.: Boxwood Press.

Lie, J. 1998. Evolution of individuality, fixed size model. Unpublished manuscript.

Lohrmann, R. 1975. Formation of nucleoside 5'-polyphosphates from nucleotide and trimetaphosphate. *Journal of Molecular Evolution* 6:237–252.

Lohrmann, R. and Orgel, L. E. 1980. Efficient catalysis of polycytitylic acid–directed oligoguanylate formation by $PB^{2+}$. *Journal of Molecular Biology* 142:555–567.

Lohrmann, R., Bridson, P. K., and Orgel, L. E. 1980. Efficient metal-ion catalyzed template-directed oligonucliotide synthesis. *Science* 208:1464–1465.

Long, A. and Michod, R. E. 1995. Origin of sex for error repair. I. Sex, diploidy, and haploidy. *Theoretical Population Biology* 47:18–55.

MacArthur, R. H. 1972. *Geographical Ecology. Patterns in the Distribution of Species.* New York: Harper and Row.

Manicacci, D., Couvet, D., Belhassen, E., Gouyon, P. H., and Atlan, A. 1996. Founder effects and sex ratio in the gynodioecious *Thymus vulgaris* L. *Molecular Ecology* 5:63–72.

Marchant, J. 1975. *Alfred Russell Wallace Letters and Reminiscences.* New York: Arno Press.

Margulis, L. 1970. *Origin of Eukaryotic Cells.* New Haven: Yale University Press.

———. 1981. *Symbiosis in Cell Evolution.* San Francisco: W. H. Freeman.

———. 1993. *Symbiosis in Cell Evolution, Microbial Communities in the Archean and Proterozoic Eons.* 2nd ed. New York: W. H. Freeman.

Marrow, P., Johnstone, R. A., and Hurst, L. D. 1996. Riding the evolutionary streetcar: where population genetics and game theory meet. *Trends in Ecology and Evolution* 11:445–446.

Maurice, S., Charlesworth, D., Desfeux, C., Couvet, D., and Gouyon, P. H. 1993. The evolution of gender in hermaphrodites of gynodioecious populations with nucleo-cytoplasmic male-sterility. *Proceedings of the Royal Society of London B, Biological Sciences* 251:253–261.

Maurice, S., Belhassen, E., Couvet, D., and Gouyon, P. H. 1994. Evolution of dioecy: Can nuclear-cytoplasmic interactions select for maleness? *Heredity* 73:346–354.

Maynard Smith, J. 1978. *The Evolution of Sex.* London: Cambridge University Press.

———. 1981. Will a sexual population evolve to an ESS? *American Naturalist* 117:1015–1018.

―――. 1982. *Evolution and the Theory of Games.* London: Cambridge University Press.

―――. 1988. Evolutionary progress and levels of selection. Pages 219–230 in *Evolutionary Progress,* edited by M. H. Nitecki. Chicago: University of Chicago Press.

―――. 1990. Models of a dual inheritance system. *Journal of Theoretical Biology* 143:41–53.

―――. 1991a. A Darwinian view of symbiosis. Pages 26–39 in *Symbiosis as a Source of Evolutionary Innovation,* edited by L. Margulis and R. Fester. Cambridge, Mass.: MIT Press.

―――. 1991b. Byerly and Michod on fitness. *Biology and Philosophy* 6:37.

Maynard Smith, J. and Price, G. R. 1973. The logic of animal conflict. *Nature (London)* 246:15–18.

Maynard Smith, J. and Szathmáry, E. 1993. The origin of chromosomes I. Selection for linkage. *Journal of Theoretical Biology* 164:437–446.

―――. 1995. *The Major Transitions in Evolution.* San Francisco: W. H. Freeman.

McClain, M. E. and Spendlove, R. S. 1966. Multiplicity reactivation of reovirus particles after exposure to ultraviolet light. *Journal of Bacteriology* 92:1422–1429.

Metz, J. A. J., Geritz, S. A. H., Meszena, G., Jacobs, F. J. A., and van Heerwaarden, J. S. 1996. Adaptive dynamics: a geometrical study of the consequences of nearly faithful reproduction. Pages 183–231 in *Stochastic and Spatial Structures of Dynamical Systems,* edited by S. J. van Strien and S. M. Verduyn Lunel. Amsterdam: North Holland.

Metz, J. A. J., Nisbet, R. M., and Geritz, S. A. H. 1992. How should we define "fitness" for general ecological scenarios? *Trends in Ecology and Evolution* 7:198–202.

Michelson, S., Miller, B. E., Glicksman, A. S., and Leith, J. T. 1987. Tumor microecology and competitive interactions. *Journal of Theoretical Biology* 128:233–246.

Michod, R. E. 1979. Genetical aspects of kin selection: effects of inbreeding. *Journal of Theoretical Biology* 81:223–233.

―――. 1980. Evolution of interactions in family structured populations: mixed mating models. *Genetics* 96:275–296.

―――. 1981. Positive heuristics in evolutionary biology. *British Journal of the Philosophy of Science* 32:1–36.

―――. 1982. The theory of kin selection. *Annual Review of Ecology and Systematics* 13:23–55.

―――. 1983a. Book review of *The Extended Phenotype: The Gene as the Unit of Selection* by Richard Dawkins. *American Scientist* 71:525–526.

―――. 1983b. Population biology of the first replicators: on the origin of the genotype, phenotype and organism. *American Zoologist* 23:5–14.

―――. 1984. Constraints on adaptation with special reference to social behavior. Pages 253–278 in *The New Ecology: Novel Approaches to Interactive Systems,* edited by P. W. Price, C. N. Slobodchikoff, and W. S. Gaud. New York: John Wiley & Sons.

―――. 1986. On fitness and adaptedness and their role of evolutionary explanation. *Journal of History of Biology* 19:289–302.

246

——. 1989. Darwinian selection in the brain. Book review of *Neural Darwinism, The Theory of Neuronal Group Selection* by G. M. Edelman. *Evolution* 43:694–696.

——. 1990. Neural Edelmanism. Letter to the editor, Reply to Francis Crick. *Trends in Neuroscience* 13:12–13.

——. 1991. Sex and evolution. Pages 285–320 in *1990 Lectures in Complex Systems, SFI Studies in the Sciences of Complexity*, Lecture Volume III, edited by L. Nadel and D. Stein. Reading, Mass.: Addison-Wesley.

——. 1995. *Eros and Evolution: A Natural Philosophy of Sex*. Reading, Mass.: Addison-Wesley.

——. 1996. Cooperation and conflict in the evolution of individuality. II. Conflict mediation. *Proceedings of the Royal Society of London B, Biological Sciences* 263:813–822.

——. 1997a. Cooperation and conflict in the evolution of individuality. I. Multi-level selection of the organism. *American Naturalist* 149:607–645.

——. 1997b. Evolution of the individual. *American Naturalist* 150:S5–S21.

——. 1997c. What good is sex? *The Sciences* 37:42–46.

——. 1998. Origin of sex for error repair. III. Selfish sex. *Theoretical Population Biology* 53:60–74.

Michod, R. E. and Abugov, R. 1980. Adaptive topography in family structured models of kin selection. *Science* 210:667–669.

Michod, R. E. and Anderson, W. W. 1979. Measures of genetic relationship and the concept of inclusive fitness. *American Naturalist* 114:637–647.

Michod, R. E. and Hamilton, W. D. 1980. Coefficients of relatedness in sociobiology. *Nature (London)* 288:694–697.

Michod, R. E. and Hasson, O. 1990. On the evolution of reliable indicators of fitness. *American Naturalist* 135:788–808.

Michod, R. E. and Levin, B. R., ed. 1988. *Evolution of Sex: An Examination of Current Ideas*. Sunderland, Mass.: Sinauer Associates.

Michod, R. E. and Long, A. 1995. Origin of sex for error repair. II. Rarity and extreme environments. *Theoretical Population Biology* 47:56–81.

Michod, R. E. and Roze, D. 1997. Transitions in individuality. *Proceedings of the Royal Society of London B, Biological Sciences* 264:853–857.

——. 1998. Cooperation and conflict in the evolution of individuality. III. Transitions in the unit of fitness. Pages 47–91 in *Mathematical and Computational Biology: Computational Morphogenesis, Hierarchical Complexity, and Digital Evolution*, edited by C. L. Nehaniv. Lectures on Mathematics in the Life Sciences, vol. 26. American Mathematical Society. (Actual publication year was 1999; available at http://eebweb.arizona.edu/michod)

Michod, R. E. and Sanderson, M. J. 1985. Behavioural structure and the evolution of social behaviour. Pages 95–104 in *Evolution—Essays in Honour of John Maynard Smith*, edited by J. J. Greenwood and M. Slatkin. Cambridge, England: Cambridge University Press.

Miller, S. L. 1987. Which organic compounds could have occurred on the prebiotic earth? *Cold Spring Harbor Symposium of Quantitative Biology* LII:9–16.

Mills, S. and Beatty, J. 1979. The propensity interpretation of fitness. *Philosophy of Science* 46:263–288.

Miyaki, M., Konishi, M., Kikuchi-Yanoshita, R., Enomoto, M., Igari, T., Tanaka, K., Mraoka, M., Takahashi, H., and Amada, Y. 1994. Characteristics of somatic mutation of the adenomatous polyposis coli gene in colorectal tumors. *Cancer Research* 54:3011–3020.

Møller, A. P. and Pomiankowski, A. 1993. Fluctuating asymmetry and sexual selection. *Genetica* 89:267–279.

Morrell, V. 1996. Genes versus teams. *Science* 273:739–740.

Muller, H. J. 1932. Some genetic aspects of sex. *American Naturalist* 66:118–138.

———. 1964. The relation of recombination to mutational advance. *Mutation Research* 1:2–9.

Nielsen, K. V., Madsen, M. W., and Briand, P. 1994. In vitro karyotype evolution and cytogenetic instability in the non-tumorigenic human breast epithelial cell line HMT-3522. *Cancer Genetics and Cytogenetics* 78:189–99.

Noest, A. J. 1997. Instability of the sexual continuum. *Proceedings of the Royal Society of London B, Biological Sciences* 264 (1386):1389–1393.

Nowak, M. A. 1990a. An evolutionaily stable strategy may be inaccessible. *Journal of Theoretical Biology* 142:237–241.

———. 1990b. Stochastic strategies in the Prisoner's Dilemma. *Theoretical Population Biology* 38:93–112.

Nowak, M. A. and May, R. M. 1992. Evolutionary games and spatial chaos. *Nature (London)* 359:826–829.

———. 1993. The spatial dilemmas of evolution. *International Journal of Bifurcations and Chaos* 3:35–78.

Nowak, M. A. and Sigmund, K. 1992. Tit for tat in heterogeneous populations. *Nature (London)* 355:250–253.

———. 1993. A strategy of win-stay, lose-shift that outperforms tit-for-tat in the Prisoner's Dilemma game. *Nature (London)* 364:56–58.

Nowell, P. C. 1976. The clonal evolution of tumor cell populations. *Science* 194:23–28.

Nur, N. 1984. Fitness, population growth rate and natural selection. *Oikos* 42:413–415.

———. 1987. Population growth rate and the measurement of fitness: a critical reflection. *Oikos* 48:338–341.

Oparin, A. I. 1938. *Origin of Life.* New York: Dover.

———. 1965. The pathways of the primary development of metabolism and artificial modeling in coacervate drops. Pages 331–346 in *The Origins of Prebiological Systems,* edited by S. W. Fox. New York: Academic Press.

———. 1968. *Genesis and Evolutionary Development of Life.* New York: Academic Press.

Orgel, L. E. 1986. RNA catalysis and the origins of life. *Journal of Theoretical Biology* 123:127–149.

———. 1987. Evolution of the genetic apparatus: a review. *Cold Spring Harbor Symposium on Quantitative Biology* LII:9–16.

———. 1992. Molecular replication. *Nature (London)* 358:203–209.

Orlove, M. J. 1975. A model of kin selection not involving coefficients of relationship. *Journal of Theoretical Biology* 49:289–310.

Orlove, M. J. and Wood, C. L. 1978. Coefficients of relationship and coefficients of relatedness in kin selection: a covariance form for the rho formula. *Journal of Theoretical Biology* 73:679–686.

Otto, S. P. and Orive, M. E. 1995. Evolutionary consequences of mutation and selection within an individual. *Genetics* 141:1173–1187.

Palmer, A. R. and Strobeck, C. 1986. Fluctuating asymmetry: measurement, analysis, patterns. *Annual Review of Ecology and Systematics* 17:391–421.

Parker, G. A. and Maynard Smith, J. 1990. Optimality theory in evolutionary biology. *Nature (London)* 348:27–33.

Pianka, E. R. 1978. *Evolutionary Ecology,* 2d ed. New York: Harper & Row.

Price, G. R. 1970. Selection and covariance. *Nature (London)* 227:529–531.

———. 1972a. Extension of covariance selection mathematics. *Annals of Human Genetics* 35:485–490.

———. 1972b. Fisher's "fundamental theorem" made clear. *Annals of Human Genetics* 36:129–140.

———. 1995. The nature of selection. *Journal of Theoretical Biology* 175:389–396.

Prigogine, I. 1980. *From Being to Becoming.* San Francisco: W. H. Freeman.

Prigogine, I., Nicholis, G., and Babloyantz, A. 1972a. Thermodynamics of evolution, II. *Physics Today* 25 (12):38–44.

———. 1972b. Thermodynamics of evolution, I. *Physics Today* 25 (11):23–28.

Prout, T. 1965. The estimation of fitness from genotypic frequencies. *Evolution* 19:546–551.

———. 1969. The estimation of fitness from population data. *Genetics* 63:949–967.

———. 1971a. The estimation of fitness components and population prediction in *Drosophila.* II. Population prediction. *Genetics* 68:151–167.

———. 1971b. The relation between fitness components and population prediction in *Drosophila.* I. The estimation of fitness components. *Genetics* 68:127–149.

Provine, W. B. 1971. *The Origins of Theoretical Population Genetics.* Chicago: University of Chicago Press.

———. 1977. Role of mathematical population geneticists in the evolutionary synthesis of the 1930's and 40's. Pages 2–30 in *Mathematical Models in Biological Discovery,* edited by D. L. Solomon and C. Walter. Berlin: Springer-Verlag.

———. 1986. *Sewall Wright and Evolutionary Biology.* Chicago and London: University of Chicago Press.

Quenette, P. Y. and Gerard, J. F. 1992. Does frequency-dependent selection optimize fitness? *Journal of Theoretical Biology* 159:381–385.

Ramel, C. 1992. Evolutionary aspects of human cancer. *Pharmacogenetics* 2:344–349.

Rand, D. A., Wilson, H. B., and McGlade, J. M. 1994. Dynamics and evolution: evolutionarily stable attractors, invasion exponents and phenotype dynamics. *Transactions of the Royal Society of London B* 343:261–283.

REFERENCES

Ransick, A., Cameron, R. A., and Davidson, E. H. 1996. Postembryonic segregation of the germ line in sea urchins in relation to indirect development. *Proceedings of the National Academy of Sciences* 93:6759–6763.

Rebek, J. 1994. Synthetic self-replicating molecules. *Scientific American* 271:48–55.

Robertson, A. 1966. A mathematical model of the culling process in daily cattle. *Animal Productions* 8:95–108.

———. 1968. The spectrum of genetic variation. Pages 5–16 in *Population biology and evolution,* edited by R. C. Lewontin. Syracuse, N.Y.: Syracuse University Press.

Rosenberg, A. 1978. Supervenience of biological concepts. *Philosophy of Science* 45:368–386.

———. 1983. Fitness. *Journal of Philosophy* 80:457–473.

———. 1985. *The Structure of Biological Science.* New York: Cambridge University Press.

———. 1991. Adequacy criteria for a theory of fitness. *Biology and Philosophy* 6:38–41.

Roughgarden, J. 1979. *Theory of Population Genetics and Evolutionary Ecology: An Introduction.* New York: Macmillan Publishing.

Ruse, M. 1973. *The Philosophy of Biology.* London: Hutchinson University Library.

———. 1988. *Philosophy of Biology Today.* Albany, N.Y.: SUNY Press.

Schuster, P. 1980. Prebiotic evolution. Pages 15–87 in *Biochemical Evolution,* edited by H. Gutfreund. Cambridge, England: Cambridge University Press.

Shibata, D., Peinado, M. A., Ionov, Y., Malkhosyan, S., and Perucho, M. 1994. Genomic instability in repeated sequences is an early somatic event in colorectal tumorigenesis that persists after transformation. *Nature Genetics* 6:273–281.

Sober, E. 1984. *The Nature of Selection.* Cambridge, Mass.: MIT Press.

———. 1993. *Philosophy of Biology.* Boulder, Colo.: Westview Press.

Soule, M. E. 1979. Heterozygosity and developmental stability: another look. *Evolution* 33:396–401.

Stearns, S. C. 1992. *The Evolution of Life Histories.* Oxford: Oxford University Press.

———, ed. 1987. *The Evolution of Sex and Its Consequences.* Basel: Birkhauser Verlag.

Stenseth, N. C. 1984. Fitness, population growth rate and evolution in plant-grazer systems: a reply to Nur. *Oikos* 42:414–415.

Stewart, R. N. 1978. Ontogeny of the primary body in chimeral forms of higher plants. Pages 131–160 in *The Clonal Basis of Development,* edited by W. Subtelny and I. M. Sussex. New York: Academic Press.

Stewart, R. N., Meyer, F. G., and Dermen, H. 1972. Camellia + "Daisy Eagleson," a graft chimera of *Camellia sasanqua* and *C. japonica. American Journal of Botany* 59:515–524.

Szathmáry, E. 1991. Population dynamics and selection. *Trends in Ecology and Evolution* 6:366–370.

Szathmáry, E. and Maynard Smith, J. 1995. The major evolutionary transitions. *Nature (London)* 374:227–232.

————. 1997. From replicators to reproducers: the first major transitions leading to life. *Journal of Theoretical Biology* 187:555–571.

Talbot, C. C. J., Avramopoulos, D., Gerken, S., Chakravarti, A., Armour, J. A., Matsunami, N., White, R., and Antonarakis, S. E. 1995. The tetranucleotide repeat polymorphism D21S1245 demonstrates hypermutability in germline and somatic cells. *Human Molecular Genetics* 4:1193–1199.

Taylor, P. D. and Jonker, L. B. 1978. Evolutionarily stable strategies and game dynamics. *Mathematical Biosciences* 40:145–156.

Temin, H. M. 1988. Evolution of cancer genes as a mutation-driven process. *Cancer Research* 48:1697–1701.

Templeton, A. 1982. Adaptation and the integration of evolutionary forces. Pages 15–31 in *Perspectives on Evolution,* edited by R. Milkman. Sunderland, Mass.: Sinauer Associates.

Thoday, J. M. 1953. Components of fitness. *Symposium of the Society for Experimental Biology* 7:96–113.

Towe, R. M. 1981. Environmental conditions surrounding the origin and early archean evolution of life: A hypothesis. *Precambian Research* 16:1–10.

Tsiotou, A. G., Krespis, E. N., Sakorafas, G. H., and Krespi, A. E. 1995. The genetic basis of colorectal cancer—clinical implications. *European Journal of Surgical Oncology* 21:96–100.

Uyenoyama, M. K. and Feldman, M. W. 1981. On relatedness and adaptive topography in kin selection. *Theoretical Population Biology* 19:87–123.

Van Holde, K. E. 1980. The origin of life: A thermodynamic critique. Pages 31–46 in *The Origin of Life and Evolution,* edited by H. O. Halvorson and K. E. Van Holde. New York: Alan R. Liss.

Wade, M. J. 1978. Kin selection: a classical approach and a general solution. *Proceedings of the National Academy of Sciences* 75:6154–6158.

————. 1979. The evolution of social interactions by family selection. *American Naturalist* 113:399–417.

————. 1980. Kin selection: its components. *Science* 210:665–667.

————. 1985. Soft selection, hard selection, kin selection, and group selection. *American Naturalist* 125:61–73.

Weismann, A. 1890. Prof. Weismann's theory of heredity. *Nature (London)* 41:317–323.

Werren, J. H. 1997. Biology of *Wolbachia. Annual Review of Entomology* 42:587–609.

Werren, J. H., Nur, U., and Wu, C. 1988. Selfish genetic elements. *Trends in Ecology and Evolution* 3:297–302.

Whitham, T. G. and Slobodchikoff, C. N. 1981. Evolution by individuals, plant-herbivore interactions, and mosaics of genetic variability: the adaptive significance of somatic mutations in plants. *Oecologia* 49:287–292.

Williams, G. C. 1992. *Natural Selection: Domains, Levels, and Challenges.* Oxford and New York: Oxford University Press.

Wilson, D. S. 1975. A theory of group selection. *Proceedings of the National Academy of Sciences* 72:143–146.

Wilson, D. S. 1980. *The Natural Selection of Populations and Communities.* Menlo Park, Calif.: Benjamin/Cummings.

Wilson, D. S., Pollock, G. B., and Dugatkin, L. A. 1992. Can altruism evolve in purely viscous populations? *Evolutionary Ecology* 6:331–341.

Wilson, E. O. 1975. *Sociobiology: The New Synthesis.* Cambridge, Mass.: Belknap Press of Harvard University Press.

Wimsatt, W. 1980. Reductionistic research strategies and their biases in the units of selection controversy. Pages 213–259 in *Scientific Discovery.* Vol. II, *Case Studies,* edited by T. Nichols. Dordrecht: Reidel.

Woese, C. R. 1967. *The Genetic Code: The Molecular Basis for Genetic Expression.* New York: Harper Row.

———. 1980. An alternative to the Oparin view of the primeval sequence. Pages 77–86 in *The Origins of Life and Evolution,* edited by H. O. Halvorson and K. E. Van Holde. New York: Alan R. Liss.

Wright, S. 1931. Evolution in Mendelian populations. *Genetics* 16:97–159.

———. 1951. The genetical structure of populations. *Annals of Eugenics* 15:323–354.

———. 1969. *Evolution and the Genetics of Populations.* Vol. 2, *The Theory of Gene Frequencies.* Chicago: University of Chicago Press.

———. 1977. *Evolution and the Genetics of Populations.* Vol. 3, *Experimental Results and Evolutionary Deductions.* Chicago: University of Chicago Press.

Xiao, S., Zhang, Y., and Knoll, A. H. 1998. Three-dimensional preservation of algae and animal embryos in a Neoproterozoicphosphorite. *Nature (London)* 391:553–559.

Zeeman, E. C. 1980. Population dynamics from game theory. Pages 471–497 in *Global Theory of Dynamical Systems: Proceedings of an International Conference Held at Northwestern University, Evanston, Illinois, June 18–22, 1977,* edited by Z. Nitecki and C. Robinson. Berlin: Springer-Verlag.

# Index